International Conference on Materials Engineering and Applications (ICMEA 2018)

IOP Conference Series: Materials Science and Engineering
Volume 348

Bali, Indonesia
14 – 16 January 2018

ISBN: 978-1-5108-6264-7
ISSN: 1757-8981

Printed from e-media with permission by:

Curran Associates, Inc.
57 Morehouse Lane
Red Hook, NY 12571

Some format issues inherent in the e-media version may also appear in this print version.

Printed by Curran Associates, Inc. (2018)

For permission requests, please contact the Institute of Physics
at the address below.

Institute of Physics
Dirac House, Temple Back
Bristol BS1 6BE UK

Phone: 44 1 17 929 7481
Fax: 44 1 17 920 0979

techtracking@iop.org

Additional copies of this publication are available from:

Curran Associates, Inc.
57 Morehouse Lane
Red Hook, NY 12571 USA
Phone: 845-758-0400
Fax: 845-758-2633
Email: curran@proceedings.com
Web: www.proceedings.com

Table of contents

Volume 348

International Conference on Materials Engineering and Applications

14–16 January 2018, Bali, Indonesia

Accepted papers received: 9 April 2018
Published online: 3 May 2018

Preface

Papers

Nanomaterials and Nanotechnology

Building Materials and Structures

Material Mechanics and Solid Mechanics

Material Processing and Manufacturing

IOP Conf. Series: Materials Science and Engineering **348** (2018) 011001 doi:10.1088/1757-899X/348/1/011001

PREFACE

It is appreciated you have strongly supported ICMEA 2018 which is held in Bali, Indonesia during January 14-16, 2018. 2018 International Conference on Materials Engineering and Applications (ICMEA 2018) is dedicated to issues related to materials engineering and applications.

The major goal and feature of the conference is to bring academic scientists, engineers, industry researchers together to exchange and share their experiences and research results, and discuss the practical challenges encountered and the solutions adopted. Distinguished professors have attended the conference as Keynote Speakers & Plenary Speakers & Invited Speakers. (Prof. Stephen John Pennycook, from National University of Singapore, Singapore; Prof. Takashige Omatsu, from Chiba University, Japan; Prof. Tjokorda Gde Tirta Nindhia, from Udayana University, Indonesia; Prof. Jae-Jin Shim, from Yeungnam University, South Korea)

The conference has four sessions (Topic: Material Processing and Surface Science; Building Materials and Solid Mechanics; Chemical Engineering and Petrochemistry; Material Physics and Electronic Information Technology). On January 16, 2018, there was an academic visit in Udayana University. Participants communicated with other researchers in the university. It is a golden opportunity for the students, researchers and engineers to interact with the experts and specialists to get their advice or consultation on technical matters, sales and marketing strategies.

This proceeding presents a selection from papers submitted to the conference from universities, research institutes and industries. All of the papers were subjected to peer-review by conference committee members and international reviewers. The papers selected depended on their quality and their relevancy to the conference. The volume tends to present to the readers the recent advances in the field of materials science and engineering, measuring methods and applications etc.

We would like to thank all the authors who have contributed to this volume and also to the organizing committee, reviewers, speakers, co-chairpersons, sponsors and all the conference participants for their strong support to ICMEA 2018.

Prof. Tjokorda Gde Tirta Nindhia
Udayana University, Indonesia
January 26, 2018

IOP Conf. Series: Materials Science and Engineering **348** (2018) 011002 doi:10.1088/1757-899X/348/1/011002

Organizing Committee

Conference Co-Chairs

Prof. Tjokorda Gde Tirta Nindhia, Udayana University, Indonesia

Prof. Stephen John Pennycook, National University of Singapore

Prof. ShaoJun Guo, Peking University, China

Prof. Takashige Omatsu, Chiba University, Japan

Program Chairs

Prof. Sheikh Akbar, The Ohio State University in Columbus, USA

Prof. Frank Otremba, Federal Institute for Material Research and Testing, Germany

Organizing Chair

Ms. Yang Dan, Hong Kong Chemical, Biological & Environmental Engineering Society, Hong Kong

International Technical Committee

Prof. Cengiz S Ozkan, University of California, USA

Prof. Yalin Dong, University of Akron, USA

Prof. Nicola Pugno, University of Trento, Italy

Prof. Guy Le Lay, Aix-Marseille Université, France

Prof. Zhongchun Chen, Tottori University, Japan

Prof. Khalil El-Hami, University of Hassan 1st, Morocco

Prof. Haiqing Yin, University of Science & Technology Beijing, China

Prof. Yanqing Yang, Northwestern Polytechnical University, China

Prof. Yadong Xu, Northwestern Polytechnical University, China

Prof. Yuanzhi Zhu, North China University of Technology, China

Prof. Yanjing Su, University of Science & Technology Beijing, China

Prof. Chul Hee Lee, Inha university, South Korea

Prof. Jae-Jin Shim, Yeungnam University, South Korea

Prof. O.L. Shanmugasundaram, K.S. Rangasamy College of Technology, India

Prof. Y. Sangappa, Mangalore University, India

Prof. N.ETHIRAJ, Dr. M.G.R Educational and Research Institute-University, India

Prof. Sumit Srivastava, Manipal University Jaipur, India

Assoc. Prof. Daniel H. C. Ong, Chinese University of Hong Kong, Hong Kong

Assoc. Prof. Chengliang Hu, Shanghai Jiao Tong University, China

Assoc. Prof. Xifeng Li, Shanghai Jiao Tong University, China

Assoc. Prof. Sha Shiong Ng, Universiti Sains Malaysia, Malaysia

Assoc. Prof. Sushil Kumar Jain, Manipal University Jaipur, India

IOP Conf. Series: Materials Science and Engineering **348** (2018) 011002 doi:10.1088/1757-899X/348/1/011002

IOP Conf. Series: Materials Science and Engineering **348** (2018) 011003 doi:10.1088/1757-899X/348/1/011003

Peer review statement

All papers published in this volume of *IOP Conference Series: Materials Science and Engineering* have been peer reviewed through processes administered by the proceedings Editors. Reviews were conducted by expert referees to the professional and scientific standards expected of a proceedings journal published by IOP Publishing.

IOP Conf. Series: Materials Science and Engineering **348** (2018) 012001 doi:10.1088/1757-899X/348/1/012001

Wear characterization of nano-hydroxyapatite with addition of titanium (HA-Ti)

R.Rosmamuhamadani[1], A.Z.O. Arawi[1], M.K.Talari[1], M.M.Mahat [1], N.N.Bonnia[1], Sabrina, M.Yahaya[1], S.Sulaiman [2], M.I.S., Ismail [2]

[1]Materials Technology Program, School of Physics & Material Studies, Faculty of Applied Sciences, Universiti Teknologi MARA Selangor, 40450 Shah Alam, Selangor, Malaysia
[2]Mechanical and Manufacturing Department, Faculty of Engineering, Universiti Putra Malaysia, 43400 Serdang, Selangor, Malaysia
E-mail: rosma614@salam.uitm.edu.my

Abstract. Hydroxyapatite (Ca_{10} $(PO_4)_6(OH)_2$, HA), is an attractive material of an inorganic compound whose chemical composition and crystallographic structures are similar to the composition of the bone. A natural source such as egg shells is composed of 94 wt. % of calcium carbonate ($CaCO_3$), which can be calcined as calcium oxide (CaO) by the calcinations process. The efficient temperature to produce CaO is 900 °C for 2 hours. The synthesis of nano-HA was done by the mixing the diammonium phosphate (DAP) and calcium hydroxide ($Ca(OH)_2$) and subjected into a microwave for 30 minutes at 1100 W irradiation power. Ball milling process was used for 30 minutes to mix the nano-HA with different compositions of titanium. These were pressed to form pallets by hand hydraulic pump (force= 2300 psi). The pallets then were sintered at 1200 °C with the heating rate of 3 °C/min for 2 hours. The pallets were tested by several mechanical testing including hardness, compression strength and wear. From the results, HA-25wt. %Ti composite gave the highest hardness, compression and coefficient of friction for wear test values which were 89.6 Hv, 82.5MPa and 0.76µ respectively. It showed that by adding Ti to nano-HA, the mechanical properties of nano-HA could be enhanced. The microstructure analyses by optical micrograph showed that nano-HA-Ti particles displayed shape likes needle morphology. The particles showed the high tendency to form the agglomerations.

1. Introduction

Hydroxyapatite, HA ($Ca_{10}(PO_4)_6$ $(OH)_2$), is an inorganic, insoluble salt [1]. It belongs to a family of calcium phosphates ($CaPO_4$) and is one of the most stable crystal structures of the $CaPO_4$ [2]. This inorganic salt has a similar chemical composition to that of human bones and teeth [3], and may consequently serve as a useful coating for human implantations and dental affixtures [4]. Unfortunately, because of poor sinterability, HA shows low strength and toughness, especially in wet environments under physiological condition [5]. The poor mechanical properties of pure HA have limited its use in any load-bearing applications [6]. Ti materials are generally regarded to have good biocompatibility and also possess favorable properties, such as relatively low modulus, low density, and high strength [7]. By producing a composite with HA as the matrix phase, Ti will exhibit the favorable mechanical properties and excellent biocompatibility and bioactivity.
Many significant synthesis methods have been explored to prepare HA with controllable properties. Especially for nano scale HA, there are a lot of special behaviors. For example, nanocrystalline HA

powder could enhance densification and improve the fracture toughness of HA ceramics. Many methods have been developed to prepare HA nano powders including sol–gel method, hydrothermal reaction, precipitation method, mechano-chemical, emulsion technique, wet-chemical method, template method [8] and also microwave synthesis [9], as being used in this research. However, most of these methods are costly and require a strict pH control, vigorous agitation and long time for hydrolysis. Use of nanograin ceramics can also improve sinterability due to the higher surface area, and resultant nano scale grains can enhance mechanical reliability by reducing the critical flaw size. The high volume fraction of grain boundaries in nano grain compacts can increase fracture toughness as well as other mechanical properties [10].

The sintering temperature of nano-HA with the addition of Ti was studied by microwaves synthesis [11]. They found the nano-HA can be obtained purely using microwave synthesis at 1100 W. The HA powder at 1100 W showed well defined broad crystalline peak with the average of crystallizing size was 39 nm. From the results, they found the optimum sintering temperature for nano-HA-Ti, were obtained at 1200°C. It caused by the highest value of mechanical properties such as in hardness and compression tests were recorded. The Scanning Electron Micrograph (SEM) showed that increased the sintering temperature may cause a crack to nano-HA-Ti pallets.

2. Methodology

Nano-HA was extracted from eggshells. The eggshells were calcined in a furnace at 900 °C for 2 hours to extract $CaCO_3$ from calcium oxide (CaO). The CaO produced was blend using ball milling with 200 rpm for 15 minutes to get fined powder. Then, CaO had undergone hydrolysis process where water is added to change CaO to calcium hydroxide $Ca(OH)_2$. $Ca(OH)_2$ then mixed with di-ammonium phosphate solution (DAP). The solution was synthesized by using a microwave with power 1100 W for 30 minutes to get HA solution. After that, HA solution produced was dried using freeze dryer at temperature -50 °C to get HA powder. XRD used to characterize the presence of the high purity crystalline HA. HA powder then was mixed with Ti at different percentage of Ti which was 10, 20 and 30 wt. % and made into pallet by using a hydraulic press. After that, the pallets were sintered with a heating rate of 3 °C/min and soak at 1000 °C for 2 hours. Mechanical properties of the each pallet were analyzed using hardness and compression technique. The hardness of the sintered pallets HA-Ti was tested by using Hardness Vickers Tester (SHIMADZU). Compression testing was performed strength using Instron Universal Testing (5582) machine under load 100 kN with speed of 5 mm/min. The microstructures of the surface of the samples were examined using SEM with 5000X magnification.

3. Results and discussions

Hardness values for sintered Ti-HA composite with different Ti contents were shown in Table 1. It can be clearly seen that the addition of Ti contents has resulted in increased hardness values. Ti acts as a soft phase during sintering and aids in superior densification during the sintering process. Furthermore, metallic Ti acts as a bond between ceramic HA and contributed to the rigidity of the sintered samples when Ti is added to the composite. It's clearly shown that the density of sintered sample also increased with increasing of Ti contents. Superior densification and improved bonding between the ceramic grains can effectively contribute to the improved hardness of the Ti added composites compared to pure HA.

Table 1. Hardness and density of HA-Ti composites sintered at 1200 °C

% Ti	Hardness, Hv	Density
0	69.5	2.79
5	71.5	2.80
10	80.0	2.80
15	83.3	2.85
20	83.0	2.90
25	89.6	2.91

Compressive test results are tabulated in Table 2. It can be clearly seen from the table that, an addition of Ti has resulted in improved compressive strength. As discussed earlier, the addition of Ti has resulted in superior densification and improved bonding between the ceramic particles, which has resulted in improved compressive strength.

Table 2. Compressive strength of Ti added HA-Ti composites sintered at 900 °C

% Ti	Compressive Strength, MPa
0	60.0
5	65.5
10	65.0
15	78.0
20	77.5
25	82.5

Pure Ti was mixed with the microwave processed HA nanoparticles to prepare HA-Ti composites. The mixture of HA and Ti powders are pressed into green pellets and sintered at 1200 °C temperature for 2 hours. Sintered samples were subjected to wear by employing pin-on-disk apparatus (20N load). Hardened steel disks were metallographically polished and used as disks against the sintered HA-Ti composite during the wear tests.

Figure 1 shows the curve variation of coefficient of friction with sliding time during pin on disk wear test. This variation of coefficient of friction with the Ti content in the HA-Ti composite is shown in figure 2. It can be observed from the wear data that the coefficient of friction for sintered pure HA pellets is minimal compared to the Ti added pellets. Coefficient of friction for pure HA sample is 0.62 and increased to 0.66 for 5 wt.% Ti, 0.69 for 10 wt.% Ti, 0.72 for 15 wt.% Ti, 0.73 for 20 wt.% Ti and is maximum at 0.76 for 25 wt.%. It can be clearly seen that addition of Ti to the HA has resulted in increased friction between the sample pin and the rotating steel disk during the pin-on-disk wear testing.

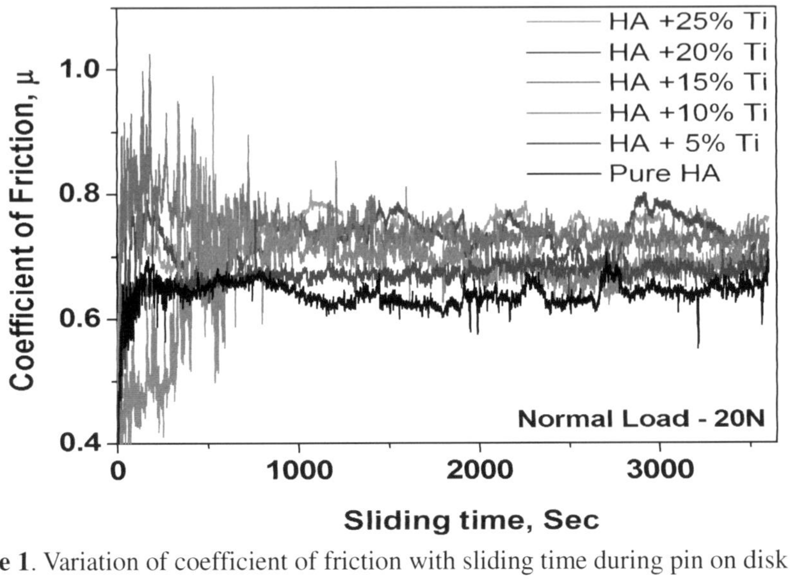

Figure 1. Variation of coefficient of friction with sliding time during pin on disk in wear test

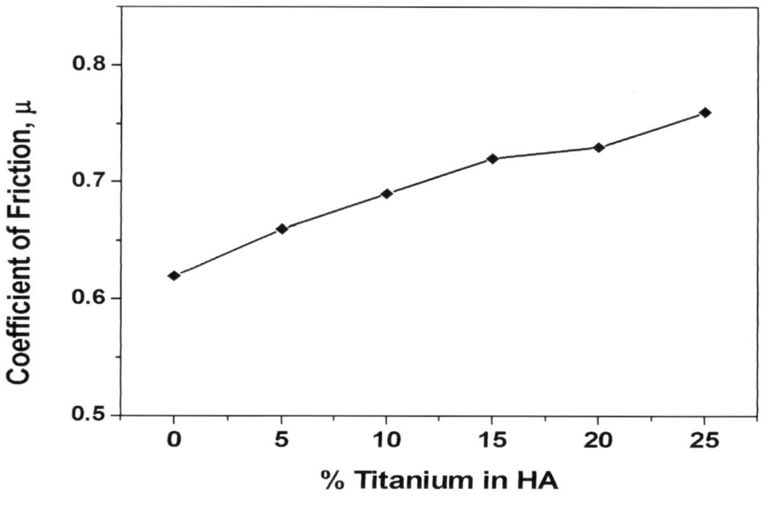

Figure 2. Graph of coefficient of friction during pin on disk wear test on the HA-Ti composites with different Ti contents.

Deep wear tracks can be observed on the surface of the wear sample. With the addition of Ti to the composite, metallic Ti being soft compared to the ceramic HA, would permit adhesion of titanium grain surface to the steel disk and follow by rupture of bonded surface. Furthermore, delaminated HA grains promote third body wear, which would typically result in deep wear track features, as observed in Figure 2 with increasing Ti content. The abrasive wear mechanism is basically the same as machining, grinding, polishing or lapping that we use for shaping materials. Two body abrasive wear occurs when one surface (usually harder than the second) cuts material away from the second, although this mechanism, very often changes to three body abrasion as the wear debris then acts as an abrasive between the two surfaces. Abrasives can act as in grinding where the abrasive is fixed relative to one surface or as in lapping where the abrasive tumbles producing a series of indentations as opposed to a scratch.

This adhesion of two surfaces under consideration and third body abrasion between surfaces during wear can result in c=increased frictional forces and result in a high coefficient of friction as observed to increase with increased Ti content. It can be seen from the Figure 3, that wear volume has decreased with the increased addition of Ti to HA-Ti composites. However, it is expected that increased coefficient of friction generally results in increased wear volume which is contrary to the current observation.

Earlier, it is stated that when two surfaces are brought together under load, asperities of the two surfaces adhere to each other. The conditions at the interface of these junctions are similar to those of a cold weld. A strong bond is formed, but without much inters diffusion of atoms and crystallization as would occur in a hot weld. During sliding, these junctions are sheared. Shearing may occur at the interface or within one of the two asperities. Most junctions shear at the interface, but occasionally shearing will occur in one of the two materials. This will result in a worn fragment being transferred from one surface to the other. It would seem that shearing should always occur at the interface since it should have the greatest weakness. This generated loose particle which in turn aid in scratching the surfaces involved in wear. All these factors increase the friction between the surfaces which can be directly correlated to the direct material loss during wear.

However, in the current work, the contrast, decreased wear volume with increased coefficient of friction was observed. This could be explained as follows. It is true that higher the friction, higher will be the material loss during wear. The above statement is true as long as wear mechanism involved remains constant with all the samples. However, as discussed earlier, the wear mechanism progressively changed from fatigue assisted delamination for pure HA to abrasive wear as Ti is added.

Furthermore, as discussed earlier, the hardness of samples increased with increased Ti content in the HA-Ti samples. Ti at the grain boundaries of HA acts as sinter aid and results in better densification during the sintering process resting in higher strength values. Sintered pure HA samples are relatively weakly bonded due to insufficient densification and can easily delaminate its grains during wear even at low frictional forces.

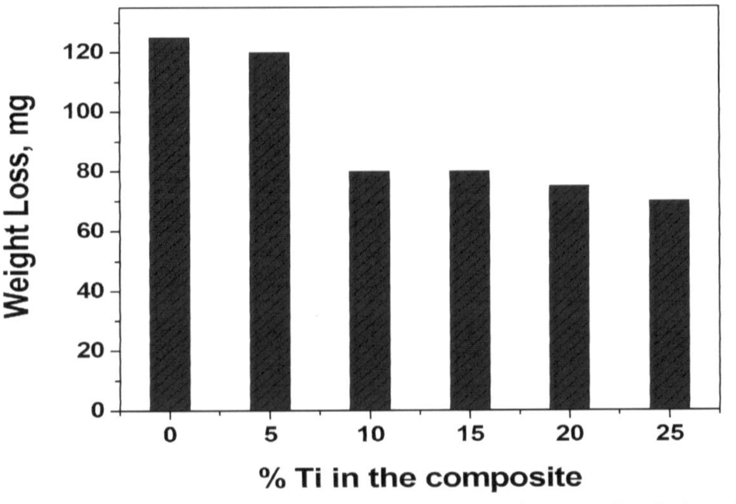

Figure 3. Effect of Ti content on the weight loss of HA-Ti composite during pin on disk wear test

Optical micrographs of worn out surfaces of composite HA samples after the pin-on-disk testing, were shown in figure 4. It can be clearly seen from the micrographs that type of wear failure has progressively changed from delamination for pure HA samples to abrasive wear as Ti content increased in the composite. Pure HA sample wear tracks have shown a surface with no considerable visual deformation and loss of material is due to delamination of small particles by the propagation of cracks. This crack propagation could be, however, inter-granular or intra-granular, assisted by fatigue of the surface during wear of the surface.

Pure HA, being a ceramic, displays a little tendency for considerable plastic deformation and as a result resists the adhesive type of wear during testing. Furthermore, no metal transfer from the steel disk to the HA sample was observed, supporting the low coefficient of friction values observed for pure HA samples. Microplastic deformation followed by subsurface crack nucleation is the preliminary stage of delamination failure during wear. These subsurface cracks propagate and lead to detachments of particles from the surface of the sample resulting in this delamination failure of surfaces underwear. No visible plastic deformation and no visible deep wear tracks are typical features of fatigue assisted delamination. With the addition of Ti to the Ti-HA composites, has resulted in the gradual transformation of delamination failure mode to combined adhesive and third body surface failure modes. Adhesive wear is produced by the formation and subsequent shearing of welded junctions between two sliding surfaces. For adhesive wear to occur it is necessary for the surfaces to be in intimate contact with each other.

Thus it can be concluded that, the addition of Ti to sintered HA-Ti composites has resulted in wear mode to change to abrasive wear from fatigue assisted delamination during wear testing. The addition of Ti has resulted in superior densification and has resulted in reduced wear volumes.

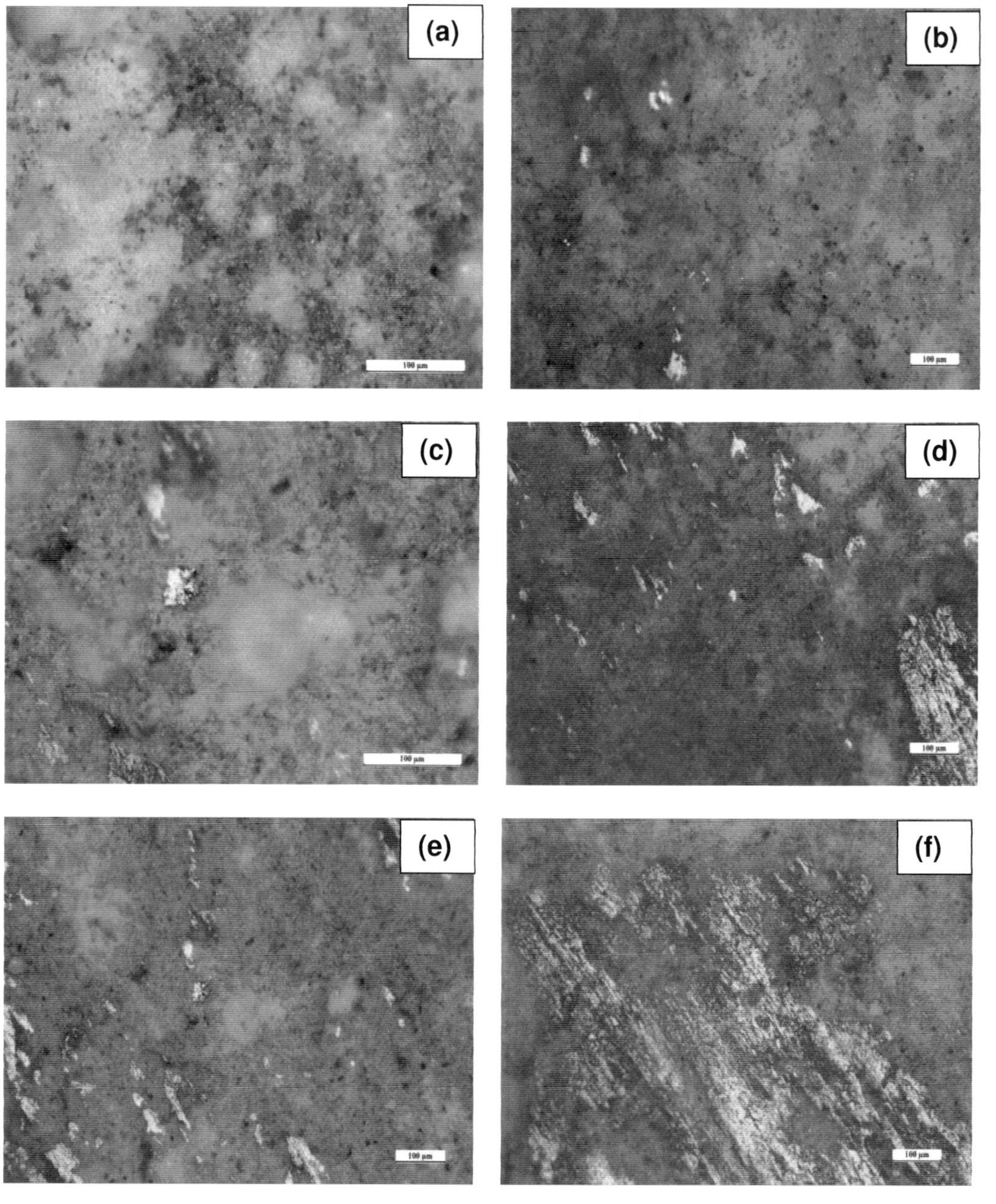

Figure 4. Optical micrographs of nano HA-Ti composite wear tracks after pin-on-disk wear tests ; a) pure HA, b) HA + 5wt.% Ti, c) HA + 10wt.% Ti ,d) HA + 15wt.% Ti ,e) HA + 20wt.% Ti and f) HA + 25wt.% Ti

4. Conclusions

In summary, it can be concluded that by sintering the sample at the temperature of 1200 °C, the highest value for hardness (290.80 Hv) and compression strength (80.04 MPa) could be obtained. Meanwhile, with the addition of 25wt. %Ti, the highest values for hardness and compression test were recorded with 89.6Hv and 82.5MPa respectively. The morphology was illustrated from optical microscope for HA-25wt. %Ti composite before sintered. It was found that the particles between nano-HA and Ti were agglomerated without any bonding. Post-sintering, the both types of particles seem to attach to each other. The size of Ti was found to be larger than nano-HA, making it more visible than nano-HA. Wear test study showed that the coefficient of HA-25 wt. % Ti composites was recorded at 0.76, while the pure HA showed at 0.62. Therefore, it can be clearly seen that the addition of Ti onto HA has resulted in the increase of friction between the sample pin and the rotating steel disk during the pin-on-disk wear testing.

5. References

[1] KJ Sato 2007, J. *Ceram. Soc. Jpn.* **115**, 124–130
[2] S Raynaud, E Champion D, Bernache-Assollant, P., Thomas 2002 *Biomaterials* **23**, 1065–1072
[3] TMG Chu, DG Orton, SJ Hollister, SE Feinberg, JW Halloran 2002 *Biomaterials* **23**, 1283–1293
[4] L El-Hammari, H Marrouna, A Laghzizila, A Saoiabia, C Rouxb, J Livageb, T Coradinb 2008 *J. Solid State Chem* **181**, 848–854
[5] A Banerjee, A Bandyopadhyay, S Bose 2008 *Materials Science and Engineering C* **27** 729-735
[6] S Bose, W Xue, A Banerjee, A Bandyopadhyay 2010 "In Handbook of nanostructured oxides for life sciences," edited by Kumar, C.S.S.R., (eds.) Weinheim: Wiley-VCH
[7] K Niespodziana, K Jurczyk, J Jakubowicz, M Jurczyk 2010 *Materials Chemistry and Physics* **123** 160–165
[8] J Klinkaewnarong, E Swatsitang, S Maensiri 2009 *Solid State Sciences* **11** 1023–1027
[9] AZO Arawi 2015 "Synthesis and Mechanical Properties of Nano-Hydroxyapatite from Egg-Shells and Effects of Nano-HA-Tiatanium Composite Using Microwave Technique" MSc Thesis, Universiti Teknologi MARA 2015.
[10] RW Rice, CC Wu, F Borchelt 1994 *J Am Ceram Soc* **77**, 2539–2553.
[11] AZO Arawi, R. Rosmamuhamadani, M.K Talari 2012 *Adv. Mater. Research* **538-541**, 2392-2395.

Acknowledgments
The authors would like to thank Faculty of Applied Sciences, Universiti Teknologi MARA (UiTM), Shah Alam and Faculty of Engineering, Universiti Putra Malaysia (UPM) for guidance, discussion and financial. This work was financially supported by FRGS from Ministry of High Education (MOHE) and Faculty of Applied Sciences, UiTM, Shah Alam.

Formation of TiO_2 based nanoribbons and the effect of post-annealing on its photocatalytic activity

D Ariyanti[1,2], S Mo'ungatonga[1], Y Li[3], and W Gao[1]

1Department of Chemical & Materials Engineering, the University of Auckland, Auckland 1142, New Zealand
2Department of Chemical Engineering, Universitas Diponegoro Semarang 50275, Indonesia
3School of Mechanical & Electrical Engineering, Henan Institute of Science and Technology, Xinxiang 453003, China
E-mail: dari293@aucklanduni.ac.nz

Abstract. The formation of TiO_2 based nanoribbons via alkaline hydrothermal methods and the post-annealing effect on its photocatalytic activity were investigated. The structure features and physical properties were characterized using X-ray diffraction (XRD), Scanning Electron Microscope (SEM) and Fourier-transform infrared (FTIR). The result shows that the morphology evolution of starting material (TiO_2 nanopowder) into TiO_2 based nanoribbons is initiated at low temperature (100°C), while the structure transformation occurred at higher temperatures (150-180°C). The process includes formation of layered titanate and dissolution-crystallisation with the ion exchange. The post-annealing process applied to the TiO_2 based nanoribbons improves its photocatalytic activity as it provides more active sites. The nanoribbon annealed at 500°C possesses high photocatalytic activity with kinetics constant of 0.082/min.

1. Introduction

Titanium oxide can be made to have many 0-2D layered titanates and 3D crystalline polymorphs, which possess special physical properties and wide range of commercial applications in pigment, photocatalysis, electronics, electrochemical and ceramic industries [1]. Different morphologies such as nanorods, nanotubes, nanosheets, nanoribbons and nanowires can be produced with different methods. Alkaline hydrothermal technique was first introduced by Kasuga *et al* in 1998 to produce TiO_2 nanotubes [2], which converts amorphous TiO_2 into nanotubes with almost 100% efficiency [2,3]. TiO_2 based nanostructures produced via alkaline hydrothermal methods possess unique physicochemical properties and can be used as catalysis, gas adsorption [4], in lithium battery [5,6] and solar cells [7]. This method can also produce TiO_2 nanoribbons, which have the specific surface area and pore volume of 20–25 m^2/g and ~0.1, cm^3/g respectively [3,4]. In this study, TiO_2 based nanoribbons were synthesised via alkaline hydrothermal methods. The effect of post-annealing on its photocatalytic activity was explored. By applying different process parameters, the evolution of morphology and transformation of crystal structure were studied. Its application on photodegradation of organic pollutants was also investigated.

2. Experimental procedures

2.1. Synthesis

Pristine TiO_2 powder (anatase, Sigma Aldrich Co. particle size < 25nm) was used as a starting material. TiO_2 based nanostructure was prepared by using an alkaline hydrothermal method. 0.2 g anatase powder was dispersed in 30 mL 10 M NaOH solution in a beaker, then immediately transferred to a Teflon-lined autoclave, which was sealed and placed in an oven at 180°C for 20 h. The products then washed first with 0.1 M HCl, followed by deionized water, and then dried at 120°C for 4 h. For the investigation of post-annealing process, TiO_2 based nanostructure were annealed at different temperature (400, 500 and 700°C) for 4 h.

2.2. Characterization
Samples was characterized by X-ray diffraction (XRD), Scanning Electron Microscope with Energy Dispersive Spectrometer (SEM-EDS, Philips XL30S FEG), and FTIR (Perkin Elmer FT-IR Spectroscopy ATR). TriStar 3000 surface area analyser was used to evaluate the Brunauer–Emmett–Teller (BET) total surface area, along with the cumulative pore volume and average pore diameter using Barrett–Joyner–Halenda (BJH) method.

2.3. Photocatalytic degradation
Photocatalytic activity was evaluated by monitoring the decomposition of Rhodamine B (RhB) in an aqueous solution under simulated solar and visible light irradiation. TiO_2 based nanoribbons (50 mg) was mixed with RhB solution (50mL, 5ppm). After stirring for 60 min in the dark to reach the adsorption equilibrium, the solution was illuminated with simulated sunlight (ULTRA VITALUX, 300W 230V E27). A sample (2 mL) was taken every particular time and filtered by 0.2 μm PVDF syringe filter prior analysis. The concentration of aqueous RhB was determined with a UV-Vis spectrophotometer (Perkin Elmer Instrument) by measuring the peak intensity at 553 nm.

3. Results and discussion

3.1. TiO_2 based nanoribbons formation via alkaline hydrothermal method
Figure 1 shows the SEM images of raw pristine TiO_2 to TiO_2 based nanoribbons via alkaline hydrothermal method. This is a simple process conducted by mixing pristine TiO_2 with 10 M NaOH solution, followed by hydrothermal process at certain temperature for 20 hours. In the relatively low temperature (100°C), the hydrothermal method abruptly changed the nanoparticle into a torn sheet-like shape (Figure 1B).

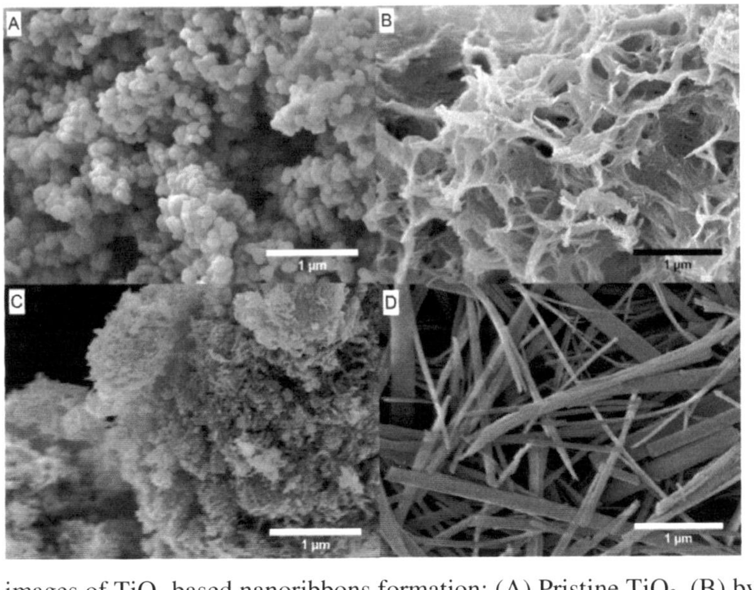

Figure 1 SEM images of TiO_2 based nanoribbons formation: (A) Pristine TiO_2, (B) by hydrothermal process at 100°C, (C) at 150°C, and (D) at 180°C.

The sheets were decorated with irregular needle-like whiskers. It was suggested that during hydrothermal process with concentrated NaOH, some Ti-O bonds from TiO_2 would be broken under high pressure inside the autoclave, resulting in the recrystallization of nanoparticles into lamellar sheet [6]. By increasing the temperature to 150°C, the whisker-like morphology evolves to nanorods of 200 nm long and 10-20 nm in diameter (Figure 1C). By setting up hydrothermal process at 180°C, regular thin nanoribbons were produced (Figure 1D), with width between 200-300 nm and length up to several microns.

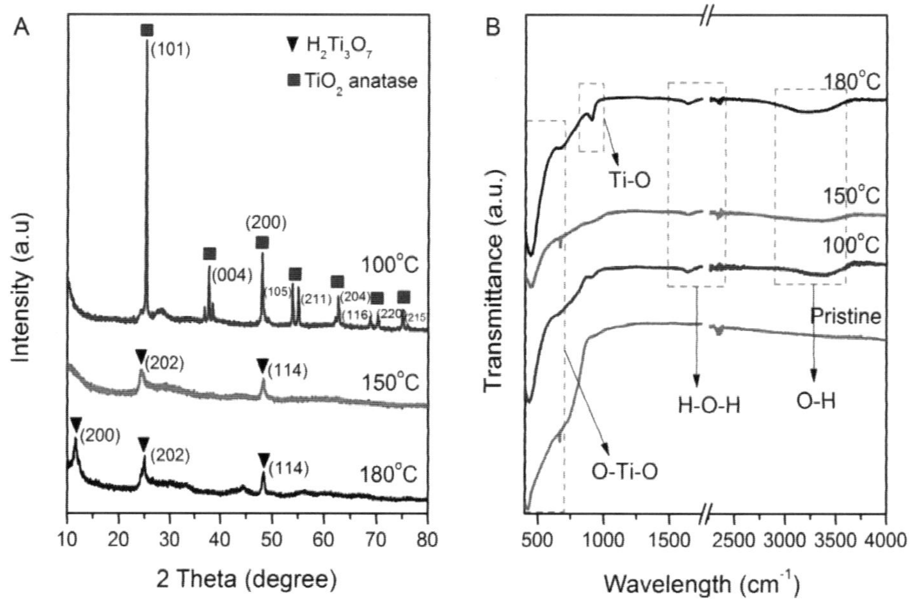

Figure 2 (A) XRD patterns and (B) FT-IR analysis of TiO_2 based nanostructures prepared at different temperatures.

XRD analysis in Figure 2A shows the crystal structure of hydrothermally treated TiO_2. Sheet-like structure formed at 100°C is confirmed to be crystalline TiO_2 anatase (JCPDS 78-2486), similar to the started material, evidence that only morphology change occurred at 100°C. The nanorods and nanoribbons formed at 150 and 180°C respectively are confirmed to be layered monoclinic structure hydrogen trititanate ($H_2Ti_3O_7$) with the peaks at 11, 24.5, 33, and 48.4° (JCPDS 47-0561). During hydrothermal process, there are several possibilities of layered protonated titanates may be produced, such as $H_2Ti_3O_7$, $H_2Ti_2O_4(OH)_2$, $H_2Ti_4O_9{\cdot}H_2O$, TiO_2-B and $H_2Ti_5O_{11}{\cdot}H_2O$, depending on the parameters used during alkaline hydrothermal treatment and acid washing [3,8].

In general, there are two reaction steps during the hydrothermal process at higher temperatures, (1) formation of layered titanate, and (2) dissolution-crystallisation along with ion exchange. The chemical reactions are suggested below [8,12]:

TiO_2 dissolved in 10 M NaOH solution for 15-20 h:

$3TiO_2 + 2NaOH \rightarrow Na_2Ti_3O_7 + H_2O$

Washing with HCl for 24 h:

$Na_2Ti_3O_7 \rightarrow 2Na^+ + Ti_3O_7^{\ 2-}$

$2Na+ + Ti_3O_7^{2-} + H^+ + Cl^- \rightarrow H^+ + Ti_3O_7^{2-} + 2Na^+ + Cl^-$

$2H^+ + Ti_3O_7^{2-} \rightarrow H_2Ti_3O_7$

Figure 2B shows the FT-IR result of TiO_2 based nanostructures prepared at different temperatures. The absorption peaks located at 443 cm^{-1} correspond to the vibrational frequency of Ti–O–Ti. Vibration mode at 908 cm^{-1} is associated with the stretching vibrations of Ti-O hydrogen titanates [3]. In addition, broad spectrum was observed at 3210 cm^{-1} and strong absorption at 1633 cm^{-1} for hydrothermally treated sample. While the former attributes to the presence of –OH groups on the surface, the latter indicates the deformation vibration of H-O-H bonds in the physically absorbed water. IR absorption in the range of 1600-3600 cm^{-1} indicates the presence of –OH groups and H-O-H molecules that are weakly bounded in different active sites of TiO_2 [9,10,11].

3.2. TiO_2 based nanoribbons formation via alkaline hydrothermal method

The photocatalytic activity of TiO_2 based nanoribbons were evaluated by the degradation of RhB under solar light simulation. Figures 3A and 3B show RhB degradation curve and kinetics constant respectively of the as produced and TiO_2 based nanoribbons annealed at different temperatures. Direct light degradation of RhB solution in the absence of photocatalyst was not detected.

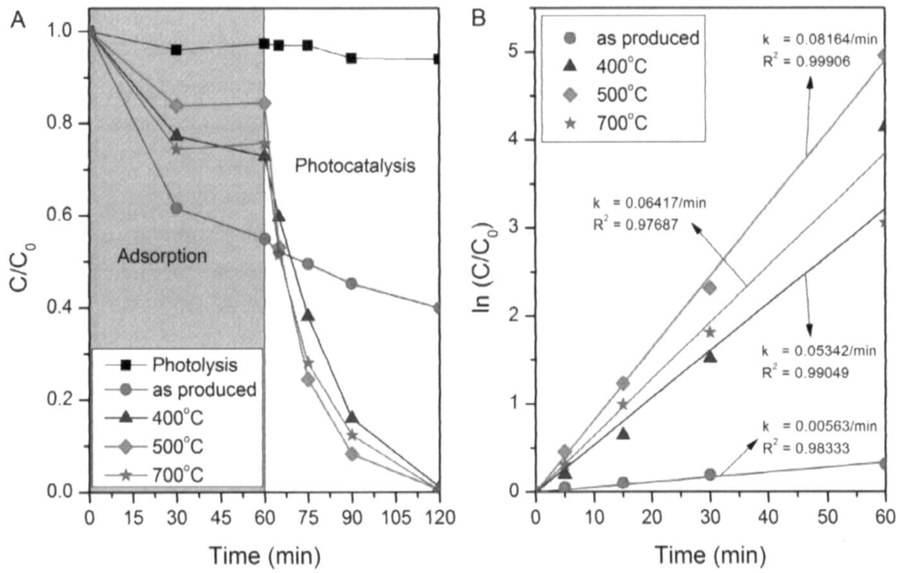

Figure 3 (A) Rhodamine B degradation profile and (B) the kinetics constant of photocatalysis using TiO_2 based nanoribbons at different annealing temperatures.

Table 1 BET and pore size distribution of TiO_2 based nanoribbons at different annealing temperatures

Samples	S_{BET} (m^2/g)	Pore volume (cm^3/g)	Pore size (nm)
As produced nanoribbon	89.805	0.384	15.896
Annealed at 500°C	37.8735	0.340	36.480
Annealed at 700°C	12.1093	0.056	23.352

After the adsorption equilibrium reached at 60 min, the as produced nanoribbon shows higher adsorption capacity than TiO_2 based nanoribbons annealed at different temperatures. The former adsorb almost 50% of RhB while the latter adsorb less than 30%. The adsorption capacities of TiO_2 based nanoribbons follow the order of annealed at 500°C > 400 and 700°C. At first, it was expected that the high adsorption capacity will result in high photocatalytic activity, as the absorbed RhB could have higher chance to be degraded in the surface of catalyst. However, the experimental result shows that the kinetics constants of TiO_2 based nanoribbons follow the order of annealed at 500°C > 400 > 700°C > as produced nanoribbons (Figure 3B). The degradation rate follows the pseudo first-order reaction kinetics; and kinetics constant is calculated by using equation $\ln(C_0/C) = kt$.

Adsorption capacity is largely influenced by pore size and pore volume of adsorbent [13,14]. Based on the surface area and pore size distribution analysis (Table 1), the as produced nanoribbons contain higher mesopores and pore volume compared to the annealed samples, which favour the adsorption of RhB molecules to the inner and narrow micropores.

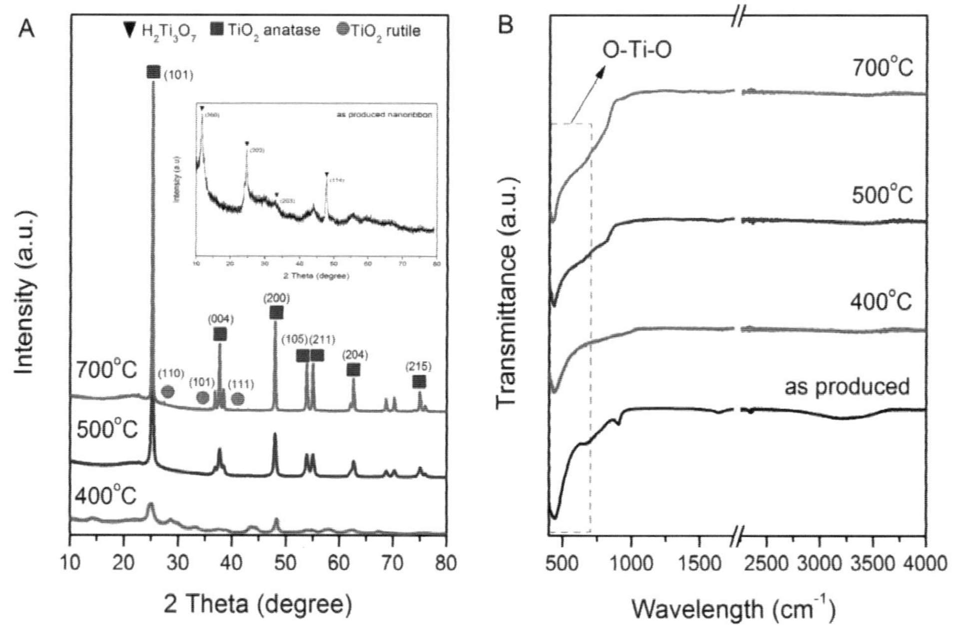

Figure 4 (A) XRD patterns and (B) FT-IR analysis of TiO_2 based nanoribbons annealed at different temperatures.

On the other hand, photocatalytic activity is mainly governed by TiO_2 crystallinity, defect structure, active site and specific surface area. Even the as produced nanoribbons possess high specific surface area; it has low number of active sites compared to TiO_2 based nanoribbons annealed at different temperatures. Figures 4A and AB shows XRD patterns and FT-IR analysis of TiO_2 based nanoribbons after annealing at different temperatures. During the annealing, the as produced nanoribbons ($H_2Ti_3O_7$) undergo rearrangement as a consequence of heating and continuous water loss [15], yielding anatase TiO_2 (JCPDS 78-2486) and small percentage of rutile TiO_2 JCPDS 78-1508 (annealed at 700°C) with the degree of crystallinity determined by the annealing temperature. Anatase structure is desirable for the photocatalytic reaction [16] due to the highly active sites.

The nanoribbon annealed at 500°C demonstrates high photocatalytic activity over nanoribbon annealed at 700°C. The former has a higher specific surface area, ~3 times compared to the latter. High specific surface area provides larger contact area for RhB molecules to the catalyst. It also associated with the surface area which provides additional active sites for absorbing light to generate charge for photocatalytic reaction. Thus the combination of TiO_2 crystallinity, number of active site and specific surface area improves photocatalytic degradation of RhB.

4. Conclusions

TiO_2 based nanoribbons produced via alkaline hydrothermal methods and post-annealing effect on its photocatalytic activity has been investigated. Two reaction steps that determine its structure during the hydrothermal process at 150-180°C include formation of layered titanate and dissolution-crystallisation along with the ion exchange. The morphology evolution is initiated at 100°C and mostly driven by the high pressure in the autoclave. The post-annealing improves the photocatalytic activity of TiO_2 based nanoribbons at a certain extent (500°C) as it can maintain the surface area and pore volume while providing higher active sites for photocatalytic reaction.

5. References

[1] Yuan Z, Colomer J, Su B. Titanium oxide nanoribbons. Chemical Physics Letters 2002 September 9,;363(3–4):362-366.
[2] Kasuga T, Hiramatsu M, Hoson A, Sekino T, Niihara K. Formation of Titanium Oxide Nanotube. Langmuir 1998 June 1,;14(12):3160-3163.
[3] Bavykin D, Friedrich J, Walsh F. Protonated Titanates and TiO2 Nanostructured Materials: Synthesis, Properties, and Applications. Adv Mater 2006;18(21):2807-2824.
[4] Umek P, Cevc P, Jesih A, Gloter A, Ewels CP, Arčon D. Impact of Structure and Morphology on Gas Adsorption of Titanate-Based Nanotubes and Nanoribbons. Chem Mater 2005 November 1,;17(24):5945-5950.
[5] Kavan L, Kalbáč M, Zukalová M, Exnar I, Lorenzen V, Nesper R, et al. Lithium Storage in Nanostructured TiO2 Made by Hydrothermal Growth. Chem Mater 2004 February 1,;16(3):477-485.
[6] Myung S, Takahashi N, Komaba S, Yoon CS, Sun Y, Amine K, et al. Nanostructured TiO2 and Its Application in Lithium-Ion Storage. Adv Funct Mater 2011 September 9,;21(17):3231-3241.
[7] Pan K, Zhang Q, Wang Q, Liu Z, Wang D, Li J, et al. The photoelectrochemical properties of dye-sensitized solar cells made with TiO2 nanoribbons and nanorods. Thin Solid Films 2007;515(7):4085-4091.
[8] Gao P, Bao D, Wang Y, Chen Y, Wang L, Yang S, et al. Epitaxial Growth Route to Crystalline TiO2 Nanobelts with Optimizable Electrochemical Performance. ACS Appl Mater Interfaces 2013 January 23,;5(2):368-373.
[9] Li G, Li L, Boerio-Goates J, Woodfield BF. High Purity Anatase TiO2 Nanocrystals: Near Room-Temperature Synthesis, Grain Growth Kinetics, and Surface Hydration Chemistry. J Am Chem Soc 2005;127(24):8659-8666.
[10] Shankar K, Basham JI, Allam NK, Varghese OK, Mor GK, Feng X, et al. Recent Advances in the Use of TiO2 Nanotube and Nanowire Arrays for Oxidative Photoelectrochemistry. J Phys Chem C 2009;113(16):6327-6359.
[11] Choudhury B, Borah B, Choudhury A. Extending Photocatalytic Activity of TiO2 Nanoparticles to Visible Region of Illumination by Doping of Cerium. Photochem Photobiol 2012;88(2):257-264.
[12] Lai CW, Bee Abd Hamid S, Tan TL, Lee WH. Rapid Formation of 1D Titanate Nanotubes Using Alkaline Hydrothermal Treatment and Its Photocatalytic Performance. Journal of Nanomaterials 2015 /05/28;2015:e145360.
[13] Hsieh C, Teng H. Influence of mesopore volume and adsorbate size on adsorption capacities of activated carbons in aqueous solutions. Carbon 2000;38(6):863-869.
[14] Dąbrowski A. Adsorption — from theory to practice. Advances in Colloid and Interface Science 2001;93(1):135-224.
[15] Rutar M, Rozman N, Pregelj M, Bittencourt C, Cerc Korošec R, Sever Škapin A, et al. Transformation of hydrogen titanate nanoribbons to TiO2 nanoribbons and the influence of the transformation strategies on the photocatalytic performance. Beilstein journal of nanotechnology 2015;6(1):831-844.
[16] Xu N, Shi Z, Fan Y, Dong J, Shi J, Hu MZ-. Effects of Particle Size of TiO2 on Photocatalytic Degradation of Methylene Blue in Aqueous Suspensions. Ind Eng Chem Res 1999;38(2):373-379.

Acknowledgments
The authors would like to thank the group of Advanced Materials and Department of Chemical and Materials Engineering, the University of Auckland for their assistances. Dessy Ariyanti also thank the Directorate General of Resources for Research, Tecnology and Higher Education, Ministry of Research, Technology, and Higher Education of Indonesia for the scholarship support.

IOP Conf. Series: Materials Science and Engineering **348** (2018) 012003 doi:10.1088/1757-899X/348/1/012003

Development of a novel nanoscratch technique for quantitative measurement of ice adhesion strength

T Loho[1] and M Dickinson[2]

1 PhD Student, Department of Chemical and Materials Engineering, University of Auckland, Auckland, NZ
2 Senior Lecturer, Department of Chemical and Materials Engineering, University of Auckland, Auckland, NZ
E-mail: tloh003@aucklanduni.ac.nz

Abstract. The mechanism for the way that ice adheres to surfaces is still not well understood. Currently there is no standard method to quantitatively measure how ice adheres to surfaces which makes ice surface studies difficult to compare. A novel quantitative lateral force adhesion measurement at the micro-nano scale for ice was created which shears micro-nano sized ice droplets (less than 3 μm in diameter and 100nm in height) using a nanoindenter. By using small ice droplets, the variables associated with bulk ice measurements were minimised which increased data repeatability compared to bulk testing. The technique provided post-testing surface scans to confirm that the ice had been removed and that measurements were of ice adhesion strength. Results show that the ice adhesion strength of a material is greatly affected by the nano-scale surface roughness of the material with rougher surfaces having higher ice adhesion strength.

1. Introduction

Ice formation and adhesion can lead to various hazards and economical losses in modern society. For example, the presence of ice under aircraft wings or boat hulls increase the vehicles weight and drag coefficient [1], impairing their performance. In areas with extreme winters, accretion of ice can lead to downed power lines and road blocks [2]. One solution to this problem is to remove the ice using mechanical or thermal methods (active de-icing), but this is energy intensive [3]. Recent literature has instead focussed on passive de-icing methods which create a surface that prevents ice from building up on the surface. This can be done by minimising the formation of condensation of water on the surface or improving the roll-off behaviour of the condensed droplets which prevent frost formation [4-6]. Other researchers have tried to prevent ice formation by reducing the mechanical adhesion shear stress of the ice-substrate interface. This includes various coatings [6,7], textured surfaces [8], or infusing porous surfaces with liquid lubricants [9].

The effectiveness of these methods is usually characterised by measuring the ice adhesion strength, defined as the shear stress required to remove ice from the surface. Various methods were developed to measure ice adhesion strength at the macro-scale, for example by mechanically shearing solid ice structures [10], using torsion to shear ice layer from a pillar [11], or using centrifugal force [12]. These methods produce different results from one another for the same surface because there are numerous variables that affect the results such as the shape and size of ice, loading rate, time of contact, sample cooling rate, sample size, and many more [11]. Table 1 shows the values of ice adhesion strength of polished stainless steel measured with different testing methods done by various researchers.

Table 1. Values of ice adhesion strength measured with different testing methods for stainless steel

Literature	Year	Surface Temperature (°C)	Ice Adhesion Strength (kPa)
Raraty, L et al [13]	1958	-10	1590 to 1960
Jellinek, H [14]	1959	-4.5	67 ± 24
Petrenko, V et al [15]	1999	-10	14 to 70
Chen, J et al [16]	2014	-15	647 ± 125
Ling, E et al [17]	2016	-15	682 ± 46
Matsumoto, K et al [18]	2016	-5	3240

Micro-nano scale ice adhesion strength testing can give more objective results of ice adhesion strength and reduce some of the variables which affect macro-scale ice adhesion strength. Matsumoto *et al* demonstrated this by measuring the true ice adhesion strength of various materials in the nano-scale using Scanning Probe Microscopy (SPM) technique and found that the ice adhesion strength of a surface is higher than what was previously measured in the macro-scale [18-20].

In this study, a new way to measure ice adhesion strength using a nanoscratch technique was investigated. This is based on a technique developed by Dickinson *et al* to measure true adhesion strength of microscopic ceramic cold-spray splats [21]. By controlling the temperature of the sample during nanoindentation test, micro-nano sized water droplets can be condensed on the surface that will then freeze into micro-nano sized ice droplets. Using nano-scratch technique as the basis, the ice droplets can be sheared off the surface and ice adhesion strength can be measured.

Stainless steel was selected as a sample due to its common use in commercial and engineering applications [22]. Xenon was used as the inert gas ion source due to its high atomic mass (131.29 g.mol−1) which leads to a higher sputtering yield than lighter noble gases. Studies have shown that nano-scale ripple patterns arise during ion implantation and the formation is dependent on the angle of incidence of the ion beam as well as the crystallographic orientation of the grains [23, 24]. In metals, the ripples have been found to vary between different grains on the surface, which has been attributed to preferential sputtering of metal grains with different crystallographic orientations on the surface [25], impurities, and surface atom diffusion [26, 27].

2. Experimental procedure

2.1. Stainless steel samples

Grade 316 stainless steel with a BA (Bright Annealed) surface finish was cut into six 10 × 12 mm pieces and then cleaned with acetone wipes. The samples were 1 mm thick. The BA surface finish is the industry standard for a smooth, bright, reflective stainless steel surface according to the Standard Specification for General Requirements for Flat-Rolled Stainless and Heat-Resisting Steel Plate, Sheet, and Strip ASTM A480/A480M standard.

Five of the samples were ion implanted with Xe^+ ions to induce atomic sputtering and create sub-micron features on the surface of the stainless steel. Xe^+ ions were extracted from a Penning ion source at an accelerating voltage of 20 kV and selected with a 90° mass analyser magnet [28]. To limit a significant increase in the temperature of the samples, the ion beam current density for all ion implantation experiment was kept within 7 to 10 $\mu A.cm^{-2}$. These ions were then implanted into the surface with different fluences ranging from 10^{16} to 10^{18} $ions.cm^{-2}$ at 0° and 45° angle of incidence to promote nano-scale features formation. During ion implantation, the pressure of the chamber was at 7×10^{-5} Pa. Table 2 shows the different ion implantation conditions applied to the sample. Simulation of the ion implantation using the Dynamic Transport of Ion in Matter (D-TRIM) software [29] yielded sputtering depths shown in Table 2.

Table 2. Different ion beam treatment conditions on the stainless steel substrate. The sputtered depths were calculated using the D-TRIM software.

#	Sample Name	Fluence (ions.cm^{-2})	Incident Angle of Ions	Sputtered Depth (nm)
1	Control	n/a	n/a	n/a
2	1E16Xe0	1×10^{16}	0°	8.56
3	5E16Xe0	5×10^{16}	0°	39.3
4	1E17Xe0	1×10^{17}	0°	86.8
5	1E17Xe45	1×10^{17}	45°	162
6	1E18Xe45	1×10^{18}	45°	1670

2.2. Topography measurement

The samples were cleaned using an ultrasonic acetone bath for 5 minutes prior to characterisation. The surface roughness of the samples was characterised using a Bruker Contour GT-K Optical Profiler to measure Root Mean Square (RMS) roughness (R_q) values. With a 5× objective magnification lens and operating under the VSI mode, the optical profiler provided a vertical resolution of 3 nm and a lateral resolution of 2 μm. Vision64 software (Bruker) was used to analyse the data and generate 3D profile images. Five 960 × 720 μm rectangular measurements at random spots on the surface of each sample were taken and averaged. SEM images were captured using an FEI Quanta 200F Scanning Electron Microscope. Secondary electrons were detected by an Everhart-Thornley Detector.

2.3. Micro-nano scale ice adhesion strength testing

Nanoscratch testing was carried out using the Hysitron TI-950 Triboindenter machine with a 3 μm fluid cell diamond cono-spherical tip (Hysitron). A fluid cell tip was used because of its extended shaft made from a ceramic-based material (Macor) that prevents thermal conduction from the piezoelectric transducer of the nanoindenter that could melt the ice droplets during testing. Humidity of the testing chamber was controlled by flowing dry N_2 gas into the testing chamber until the relative humidity was measured to be 15%.

Surface temperature was controlled using a custom-made thermoelectric cooling device with a mild steel base plate on the cold side and a water-cooled heat sink on the hot side. The schematic of this device is shown in Figure 1. The cooling stage is based on a 195 W thermoelectric Peltier cooler module (APH-199-17-10-E, European Thermodynamics). An aluminium heat sink with a heat transfer area of 2.6×10^{-3} m^2 was constructed to remove heat from the hot side of the module. The heat sink is connected to a peristaltic pump that circulates 3 mL.s^{-1} of cold water (approximately 15°C) inside it. The plastic casing and screws were 3D-printed ABS polymer designed to prevent heat loss to the surrounding environment and compress all the components together, removing any air bubbles between them.

Temperature reading is obtained from a thermistor (MP-3193, TE Technology Inc.) that reads the temperature of the mild steel base plate. To control the temperature, the Peltier module and the thermistor is connected to a Pulse Width Modulation (PWM) temperature controller (TC-48-20, TE Technology Inc.) that controls the voltage applied to the module based on the temperature reading from the thermistor. To promote thermal conduction between the heat sink, the Peltier module, the thermistor and the mild steel base plate; a layer of high-density polysynthetic silver thermal compound (Arctic Silver 5, Arctic) was applied between them. The whole stage was then placed in a safety wall made from mild steel to prevent spilling of water to the nanoindenter stage in case of leakage. The sample was fixed on the mild steel base plate using a small amount of cyanoacrylate adhesive.

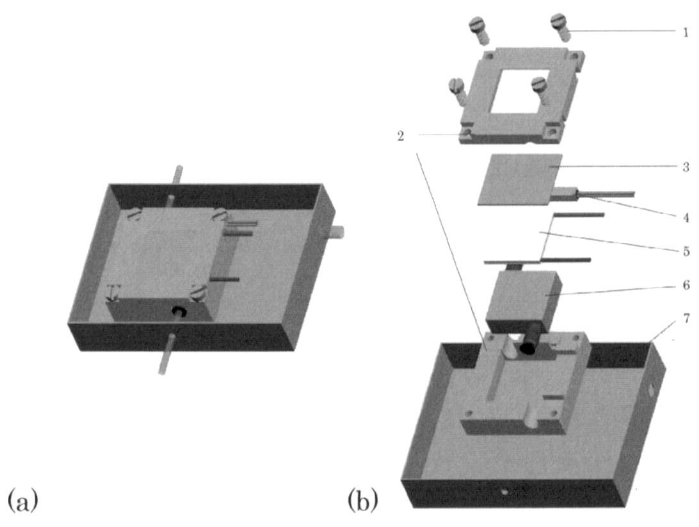

Figure 1. Custom made thermoelectric cooling device for sub-zero temperature nanoscratch testing. (a) Assembled view. (b) Exploded view of the components: (1) Plastic screws (2) Insulating plastic casing (3) Mild steel base plate (4) Thermistor (5) 195 W Thermoelectric Peltier module (6) Water cooled aluminium heat sink (7) Safety walls to prevent spilling of water in case of leakage.

The micro-nano scale ice adhesion strength testing started with performing indents on the surface to serve as a landmark, and then performing a Scanning Probe Microscopy (SPM) scan on the area by rastering a 20 × 20 μm surface of the sample at room temperature with the nanoindenter tip set at a 2 μN force setpoint. From this point onward, this image is referred to as the 'Pre-Test' SPM scan.
After the pre-test SPM scan was obtained, the nanoindenter tip was raised 100 nm above the surface and the cooling stage was turned on to reduce the surface temperature to −5◦C. This process took approximately 2 minutes before the temperature was stable, as shown in Figure 2. During this time, water droplets condensed from the surrounding air and subsequently froze into ice droplets. Due to the high thermal conductivity of the diamond tip, this process also significantly cooled the diamond cono-spherical tip of the nanoindenter and prevent melting of the ice droplets during nanoscratch testing.

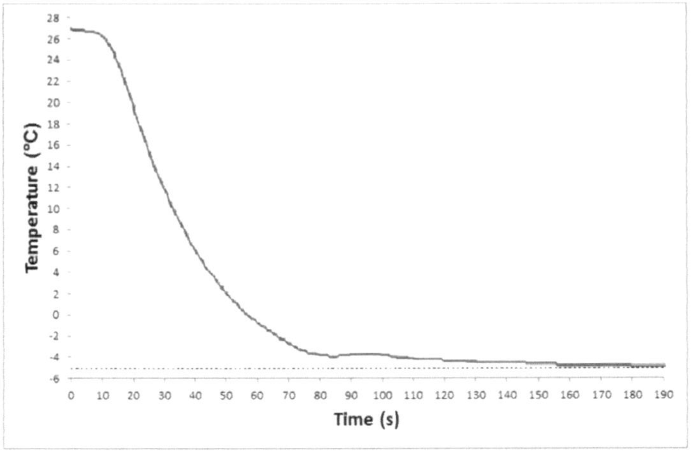

Figure 2. Temperature of the mild steel base plate over time as the cooling stage is turned on. During testing, the temperature remained stable at −5 ± 0.5°C for at least two continuous hours.

After more than two minutes had passed and the surface temperature had stabilised at −5°C, the tip was lowered into contact with the sample. Another 20 × 20 μm SPM scan of the surface was done with the nanoindenter tip set at 2 μN force setpoint at −5°C. This image is called the 'Pre-Scratch' SPM scan image. New features that were found in the Pre-Scratch image that were not in the Pre-Test image were assumed to be ice droplets that formed on the surface of the sample. A sample comparison between the Pre-Test and Pre-Scratch image of the polished stainless steel substrate is shown in Figure 3. The diameter of the ice droplet was measured using the Hysitron TriboView software to analyse the

Pre-Scratch image. Testing was done on ice droplets that are smaller than 3 μm in diameter (less than the tip's radius of curvature) to ensure the nanoscratch tip sheared the whole ice droplet and not cut through it. Once a suitable ice droplet had been found, the tip was centred above that particular ice droplet.

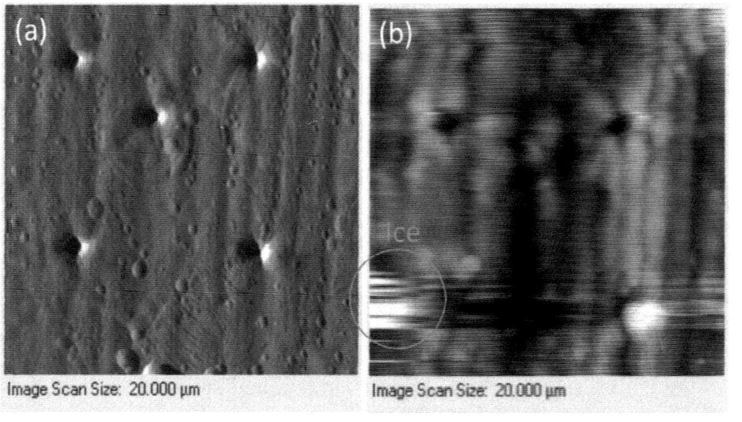

Figure 3. Gradient Forward (GF) images of the (a) Pre-Test SPM scan at room temperature and (b) Pre-Scratch SPM scan at −5°C of the polished stainless steel sample. Note that the blurriness of the Pre-Scratch image at (b) is due to the slight vibration caused by cooling water running in the heat sink. The circle on the Pre-Scratch image indicates the ice droplet that formed during the test. The tip had sliding effects at lines where the ice was found due to the slipperiness of ice droplets.

The nanoscratch test began with the tip moving 5 μm vertically to measure the tilt of the sample. The testing procedure consists of three steps, shown in the load function in Figure 4.

1. **Pre-Scan** scans the surface of the sample as the nanoindenter tip is moved 10 μm vertically with 0 μN axial force. The purpose of this step is to ensure that the ice droplet to be scratched is on the path of the nanoscratch test.
2. **Scratch** moved the tip 10 μm vertically in the other direction with 150 μN axial force to apply shear stress and remove the ice from the surface. 150 μN is selected as the axial force through trial and error as the amount of force large enough to remove the ice from the surface but not large enough to make the tip dig through the steel surface.
3. **Post-Scan** scans the surface of the sample 10 μm vertically after the nanoscratch was done to confirm removal of the ice droplet.

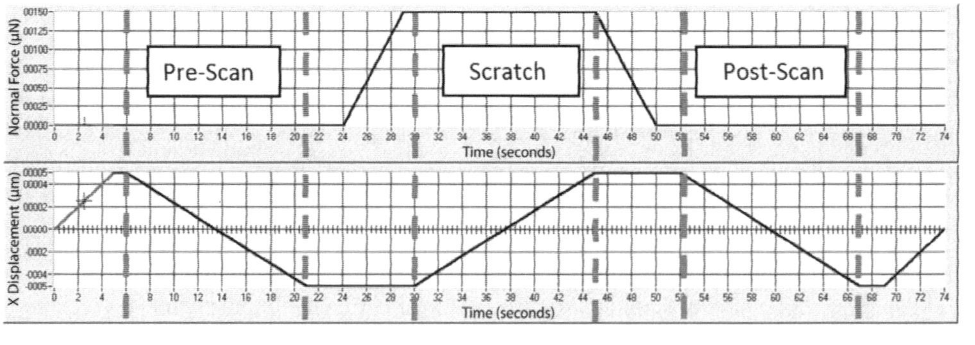

Figure 4. The load function of the nanoscratch testing procedure. The top graph shows the normal force actuated on the tip over time, while the bottom graph shows the lateral displacement of the tip over time. The load function was divided into three steps during the ice adhesion strength measurement.

3. Results and discussion

3.1. Sample topography

Optical profilometry revealed that for the first four fluences, the ion implantation process slightly increases the surface roughness of the steel. The ion implantation process creates roughness features on the surface of the steel through atomic sputtering from the xenon ions. On the 1E18Xe45 sample, it can be seen that the roughness increased significantly and a much rougher surface was created. The RMS roughness values and the 3D surface plot of the samples are shown in Figure 5.

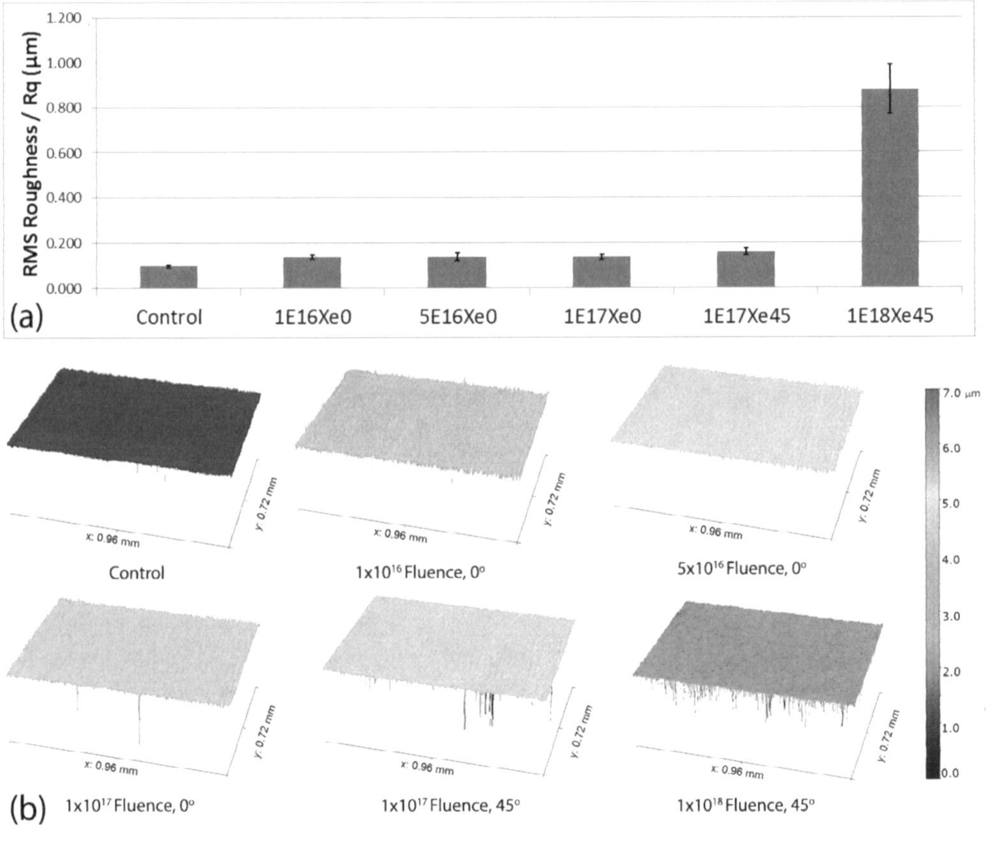

Figure 5. (a) RMS roughness and (b) 3D surface plot of the polished and ion implanted stainless steels with different fluences obtained from the optical profiler. Error bars show one standard deviation from five repeat measurements.

In order to confirm the presence of sub-micron scale features on the surface of the stainless steel, SEM images were taken at random spots on the surface of the samples, shown in Figure 6. Other than the 1E18Xe45 sample, the surfaces were seen to be relatively smooth with shallow roughness features. In the 1E18Xe45 sample, it can be seen that there are ripple patterns within the grain of the ion implanted samples, contributing to the increased roughness values measured by the optical profiler.

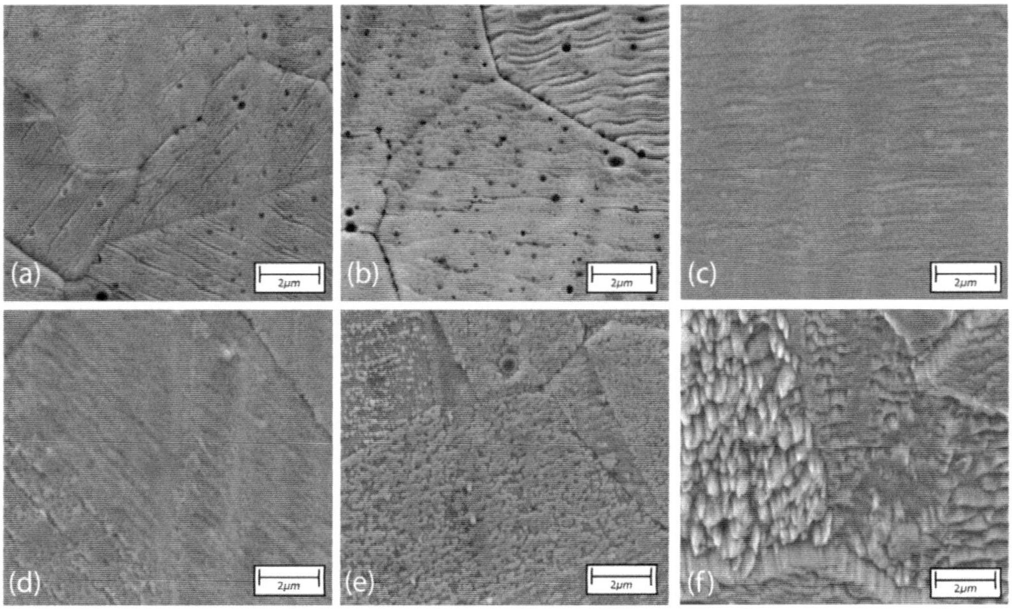

Figure 6. Secondary electrons SEM images of the (a) Control sample, (b) 1E16Xe0 sample, (c) 5E16Xe0 sample, (d) 1E17Xe0 sample, (e)1E17Xe45 sample, and (f) 1E18Xe45 sample.

3.2. Nanoscratch result analysis

To prevent overestimation of the ice droplet's diameter from the SPM scan due to the slipperiness of ice, the vertical distance of an ice droplet is measured from the Pre-Scratch SPM scan image with the Hysitron TriboView software. This is shown for the polished steel in Figure 7a and 7b and the effect is illustrated in Figure 7c. As can be seen from Figure 7b, if the horizontal distance was taken, the diameter of the ice can be overestimated by more than double the actual diameter of the ice. In this example, the ice droplet scratched was 0.74 μm in diameter and 74 nm in height. Since the ice droplet was relatively small, it can be assumed that the condensed droplet had a circular area of contact with the surface.

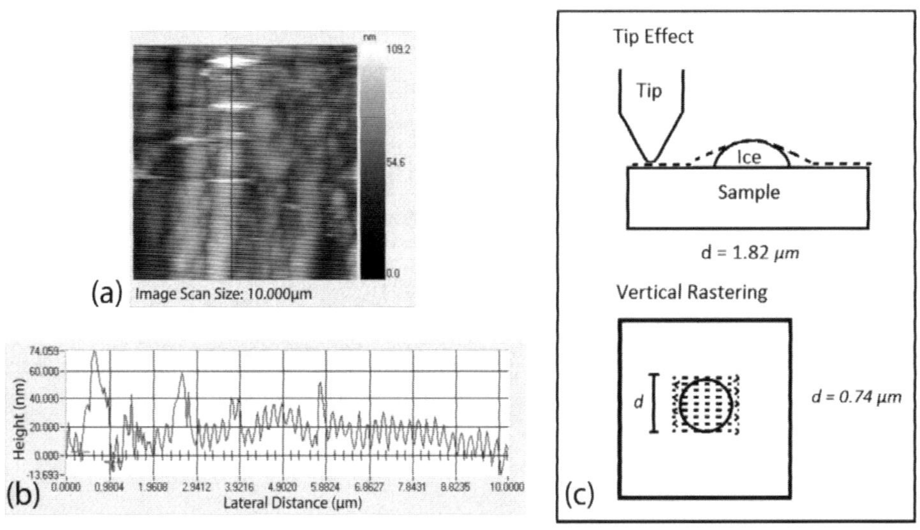

Figure 7. Determination process of an ice droplet diameter. (a) Pre-Scratch SPM image of an ice droplet; (b) Line topography profile of the SPM image shown in (a); and (c) Schematic showing why vertical distance was taken as the diameter of the ice droplet instead of the horizontal distance.

Once the size of the ice droplet was determined, a nanoscratch test according to the load function shown in Figure 4 was performed. A sample scratch result for the ice droplet measured in Figure 7a is shown in Figure 8. The analysis started by looking at the top right graph of normal displacement over time during the 'Pre-Scan' step (t = 5 to 21 s). The presence of ice was confirmed by the 'bump' (approximately 75 nm in height, similar to the ice droplet measured in Figure 7a) detected by the tip. The next step was to find the position of this bump during the 'Scratch' step of the load function. This was done by drawing a horizontal line from the position of the ice droplet at the bottom right graph of lateral displacement over time. At this position, it can be seen in the bottom left graph of lateral force over time that the tip measured a jump in force when the ice was scratched. In this example, the lateral force was measured to be 11.1 μN and this value was taken as the ice adhesion force of the sample.

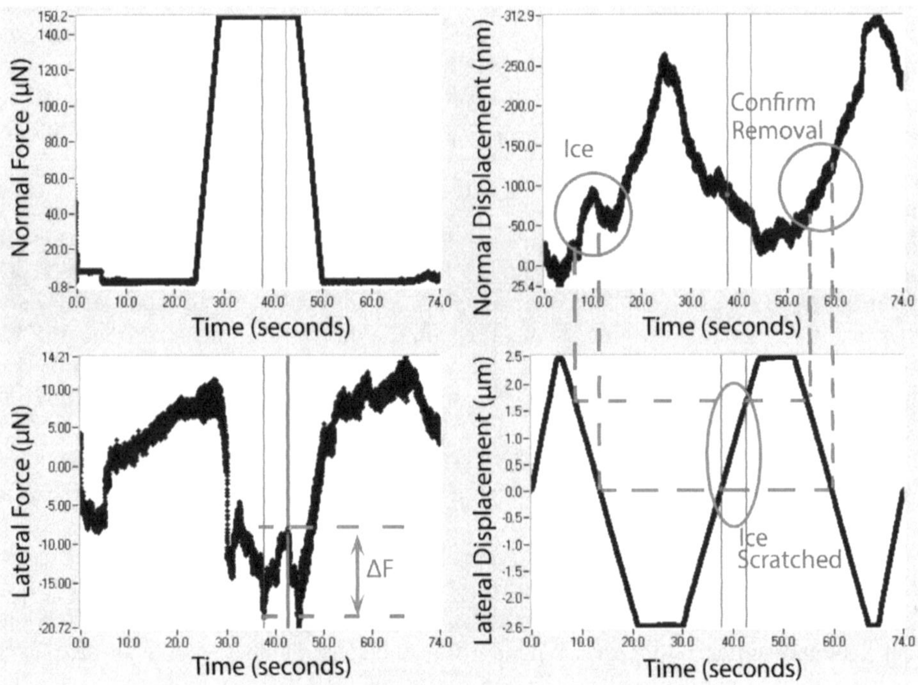

Figure 8. Scratch data obtained from the Hysitron TriboScan software during ice adhesion strength measurement after tilt correction. The four graphs shown are (top left) normal force over time (top right) normal displacement over time (bottom left) lateral force over time and (bottom right) lateral displacement over time.

The last step was to interpolate the position of the scratched ice to the 'Post-Scan' step of the load function to confirm full removal of the ice droplet from the surface, affirming that full adhesive failure had occurred. Not all of the tests resulted in full adhesive removal of the ice droplet and sometimes the Post-Scan step revealed that there was ice left behind after the scratch. These results were not considered in this study as the failure involved both adhesion and cohesion.

The ice adhesion strength of the sample was then calculated by dividing the ice adhesion force with the area of the ice droplet measured previously. In this particular example the ice adhesion strength of the polished steel was calculated to be 4.3 MPa, which is in the same order of magnitude as the nano-scale ice adhesion strength of stainless steel measured by Matsumoto *et al* [18].

3.3. Ice adhesion strength of stainless steel samples

The experiment was repeated on ten different areas of the steel samples and the results shown in Figure 9. It can be seen that the ice adhesion strength of the first five samples were not significantly different at all. This is likely because the first five samples had very similar roughness measurement and similar surface topography.

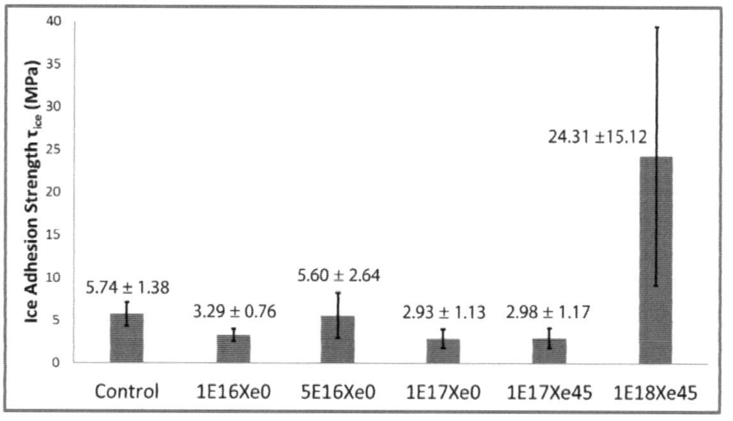

Figure 9. Mean ice adhesion strength of the polished and ion implanted stainless steels from ten repeat measurements. Error bars show one standard deviation.

Comparing this result with the ice adhesion strength values measured by different researchers in Table 1, it can be seen that this novel method yielded results in the same order of magnitude as the works of Matsumoto *et al* [18] for ice adhesion strength measurement in the nano-scale. As previously mentioned in Section 1, the true ice adhesion strength measured in the micro-nano scale was reported to be higher than the ice adhesion strength measured in the macro scale. This is thought to be caused by the absence of variables within the ice itself (e.g. impurities and air bubbles within the ice) when the size of the ice is very small. Since those variables promoted cohesive failure within the ice and could also induce pre-existing cracks, the ice adhesion strength measured in the micro-nano scale is much more dependent on the properties of the surface, such as its surface energy and topography, rather than the properties of the ice. The measured values are hence closer to the true ice adhesion strength of the surface.

It can be seen that the ice adhesion strength of the much rougher 1E18Xe45 ion implanted steel was significantly higher than the ice adhesion strength of the smoother polished steel sample and the other ion implanted samples where the roughness was not significantly modified. This is likely caused by the fact that a rougher surface had a much higher true contact area than a smooth surface for the same apparent contact area. This effect is illustrated in Figure 10. This is especially true in the case of ice because as the water droplet freezes, it expanded into the asperities, removing any air bubbles that might have been trapped between the surface of the water droplet and the solid. Because of the much higher true contact area, there was increased interaction between the solid surface and the ice droplet.

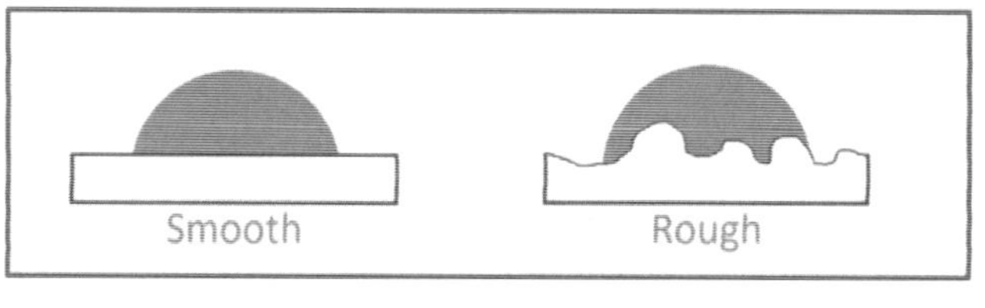

Figure 10. Schematic showing how a rough surface would have more true contact area than a smooth surface for the same apparent contact area.

4. Conclusions

A novel way to measure ice adhesion strength of materials in the micro-nano scale was developed in this study to enable a more objective and repeatable data unaffected by various factors that influence macro-scale ice adhesion failures. Moreover, by performing a post-scan of the surface after the nanoscratch procedure, the complete removal of the ice could be confirmed to ensure failure through adhesion and not cohesion which can occur in macro-scale ice adhesion tests. By performing the test on surfaces with different roughness values, it can be seen that the true ice adhesion strength is much higher for a rougher surface than a smoother surface, as a rougher surface would have much higher true contact area for the same apparent contact area.

With the development of this micro-nano scale ice adhesion strength testing method, objective comparison between ice adhesion strength reduction studies can be made easier. The advantages of this technique include the microscopic size of ice droplets, the force resolution of the nanoscratch technique that can measure μN forces, the ability to do post-scans of the surface to confirm full removal of the ice, and the stability of the substrate temperature achieved by the water-cooled thermoelectric cooler. All of this leads to a more objective result of ice adhesion strength measurement that are easy and repeatable.

5. References

[1] Civil Aviation Authority of New Zealand 2000 *Aircraft Icing Handbook* (Civil Aviation Authority (CAA))
[2] Laforte J, Allaire M and Laflamme J 1998 *Atmospheric Research* **46** 143–158
[3] Frankenstein S and Tuthill A M 2002 *Journal of Cold Regions Engineering* **16** 83–96
[4] Mishchenko L, Hatton B, Bahadur V, Taylor J A, Krupenkin T and Aizenberg J 2010 *ACS Nano* **4** 7699–7707
[5] Boinovich L B, Emelyanenko A M, Ivanov V K and Pashinin A S 2013 *ACS Applied Materials and Interfaces* **5** 2549–2554
[6] Golovin K, Kobaku S P R, Lee D H, DiLoreto E T, Mabry J M and Tuteja A 2016 *Science Advances* **2**
[7] Menini R and Farzaneh M 2011 *Journal of Adhesion Science & Technology* **25** 971–992
[8] He Y, Jiang C, Cao X, Chen J, Tian W and Yuan W 2014 *Applied Surface Science* **305** 589–595
[9] Kim P, Wong T S S, Alvarenga J, Kreder M J, Adorno-Martinez W E and Aizenberg J 2012 *ACS Nano* **6** 6569–6577
[10] Bascom W D, Cottington R L and Singleterry C R 1969 *The Journal of Adhesion* **1** 246–263
[11] Makkonen L 2012 *Journal of Adhesion Science and Technology* **26** 413–445
[12] Laforte C and Beisswenger A 2005 *International Workshop on Atmospheric Icing of Structures* **XI**
[13] Raraty L E and Tabor D 1958 *Proceedings of the Royal Society of London. Series A, Mathematical and Physical Sciences* **245** 184–201
[14] Jellinek H 1959 *Journal of colloid science* **14** 268–280
[15] Petrenko V F 1999 *Physics of ice* (Oxford: Oxford)
[16] Chen J, Luo Z, Fan Q, Lv J and Wang J 2014 *Small (Weinheim an der Bergstrasse, Germany)* **10** 4693–9
[17] Ling E J Y, Uong V, Renault-Crispo J S, Kietzig A M and Servio P 2016 *ACS Applied Materials and Interfaces* **8** 8789–8800
[18] Matsumoto K, Honda M, Minamiya K, Tsubaki D, Furudate Y and Murase M 2016 *International Journal of Refrigeration* **66** 84–92
[19] Matsumoto K and Daikoku Y 2009 *International Journal of Refrigeration* **32** 444–453
[20] Matsumoto K, Akaishi M, Teraoka Y, Inaba H and Koshizuka M 2012 *International Journal of Refrigeration* **35** 130–141
[21] Dickinson M E and Yamada M 2010 *Nanoscience and Nanotechnology Letters* **2** 348–351
[22] Cobb H M 2010 *The History of Stainless Steel* (ASM International) pp 193–228
[23] Bradley R M and Harper J M E 1988 *Journal of Vacuum Science & Technology A: Vacuum, Surfaces, and Films* **6** 2390–2395
[24] Valbusa U, Boragno C and Mongeot F B D 2002 *Journal of Physics: Condensed Matter* **14** 8153–8175
[25] Odnobokova M, Kipelova a, Belyakov a and Kaibyshev R 2014 *IOP Conference Series: Materials Science and Engineering* **63** 012060
[26] Ochoa E A, Droppa R, Basso R L O, Morales M, Cucatti S, Zagonel L F, Czerwiec T, Dos Santos M C, Figueroa C A and Alvarez F 2013 *Materials Chemistry and Physics* **143** 116–123
[27] Cucatti S, Ochoa E A, Morales M, Jr R D, Garcia J, Pinto H C and Zagonel L F 2015 4140

[28] Markwitz A and Kennedy J 2009 *International Journal of Nanotechnology* **6** 369–383
[29] Chakarov I R, Todorov S S and Karpuzov D S 1992 *The Journal of Physical Chemistry C* **69** 193–199

Acknowledgements
This work is funded by the University of Auckland Doctoral Scholarship programme with the help of Dr. Jerome Leveneur and Dr. John Kennedy from National Isotope Centre, Lower Hutt, New Zealand.

Structural and morphological study of $Zn_{0.9}Mn_{0.05}Fe_{0.05}O$ synthesized by sol-gel wet chemical precipitation route

S K Jain[1], S N Dolia[2], B L Choudhary[3] and B L Prashant[4]

1Department of Physics, Manipal School of Basic Science, Manipal University Jaipur 303007, India
2, 3 Department of Physics, University of Rajasthan, Jaipur 302005, India
4 Deptt. of Physics, A.R.S.D. College, University of Delhi, New Delhi 110021
E-mail: sushilkumar.jain@jaipur.manipal.edu

Abstract. Transition metal substituted Zinc oxide (ZnO) has drawn a great deal of attention due to its excellent properties. $Zn_{0.9}Mn_{0.05}Fe_{0.05}O$ sample synthesized was by Sol-gel wet chemical precipitation route at temperature 350°C. The crystallinity and the structure of $Zn_{0.9}Mn_{0.05}Fe_{0.05}O$ was determined by X-ray diffraction by Cu-Kα radiations operated at 40kV and 35mA in the range of 20° to 80°. The pattern gets indexed in wurtzite (hexagonal) structure with lattice constants a=b= 3.2525Å and c= 5.2071Å and approves the single phase material with no impurity. The values of particle size assessed by Debye Scherer's (DS) formula lie in the range of 13nm to 33nm indicating the nano-crystalline nature of the sample. The morphological analysis of the sample was performed by Scanning electron microscopy (SEM) and Transmission electron microscopy (TEM) measurements. The observed size of $Zn_{0.9}Mn_{0.05}Fe_{0.05}O$ nanoparticles by TEM micrograph exhibits the similar trend with the size calculated by Debye-Scherer formula. TEM image show the irregular shape of the nanoparticles and particle size lies in the range of 10-35nm. Similar to SEM image, the slight agglomeration of the nanoparticles have been observed from TEM.

1. Introduction
Semiconducting nanoparticles are of tremendous interest now-a-days for fundamental as well as applied research because the properties of the materials change significantly when the particle size reaches bulk to nano size range. Semiconducting nanoparticles are one of the key player in the industrially important materials due to size-dependent properties like alteration in the optical properties and variation of the energy gap. Zinc oxide is an excellent n-type semiconductor with a extensive direct band gap and great exciton binding energy [1, 2]. "3d" transition metal substituted ZnO material, has been of great interest for industrial applications like Optoelectronics, UV- light emitting diodes, Chemical sensors, piezoelectric devices. For Optoelectronic field, it is important to tune the optical band gap which can be done by way of doping, with various "3d" transition metal [3-7]. Doping with transition metals offers generation of dilute magnetic semiconductors (DMS) into nonmagnetic semiconductors. Dietl's reported the room temperature ferromagnetism of transition metal doped ZnO and it can be used for "spintronics" devices [8, 9]. Transition metals which contain unpaired electron or have partially filled d- states, are responsible for materials to exhibit magnetic behaviour. This property makes them for use as non-volatile switching elements [8, 9]. Over last decades, most of the research efforts were made to Mn and Co doped ZnO. Very less studies were reported for Mn and Fe doped ZnO [10]. Co-doped Mn and Fe in ZnO nanostructures may contribute the local magnetic moments and enhancement of the charge carrier density. ZnO nanoparticle can be synthesized on a

large scale at low cost by simple-solution methods, such as chemical precipitation [11] and gel combustion method [12]. The sol- gel technique permits mixing of the chemicals at microscopic level, thus reducing the possibility of undetectable impurity phase [13] also have the numerous benefit over other synthesizing processes like use of simple apparatus, low cost, environment friendly and reliability. In this paper, we report on the structural and morphological study of $Zn_{0.9}Mn_{0.05}Fe_{0.05}O$ synthesized by sol gel wet chemical precipitation method.

2. Experimental details

A basic procedure to synthesize particles was employed, on the lines of the "Co-Precipitation" method which contains dissolving a salt precursor in water to precipitate the hydroxide form with the help of a base [14-15].The focus was on preparing pure transition metal doped Zinc oxide. The Zinc chloride ($ZnCl_2$), Manganese chloride ($MnCl_2.4H_2O$), Iron chloride hexa hydrated ($FeCl_3.6H_2O$) were used without further purification. The precursors were mixed in 100 ml water in desired proportions. Solutions were stirred for half an hour with a magnetic stirrer to get homogeneous solution. Citric acid anhydrous ($C_6H_8O_7$) was added to the solution. After an hour of stirring, 10ml ethylene glycol ($C_2H_6O_2$) added and pH was maintained at 1.5 with continued stirring. Liquid ammonia was added drop wise with constant monitoring of the pH value to bring it to 7. Extra precaution was taken while adding liquid ammonia, as an overdose may render the entire solution useless for the purpose of the experiment. The solution was further stirred for another 30 minutes after adding ammonia, attaining the right pH, at a temperature of 60°C. The solution was then poured on to a watman filter paper to extract the precipitate from the sol. The precipitate (gel) was then gently scraped from the filter paper, and then washed several times with distilled water and ethyl alcohol. Washed gel was dried in hot air oven at 110°C for two hours and then cooled and ground using agate mortar and pestle, to get a fine powder. A final heating was done in a muffle furnace at 350°C for a period of two hours, before subjecting the sample to analytical measurements. Structural characterization is performed using a Panalytical 'X' pert MPD X-ray diffractometer with Cu-Kα radiation (λ= 0.15418 nm) operating at 40 kV and 35 mA for 2θ values ranging from 20° to 80°. Surface morphologies and structural details of sample have been carried out using SEM (Carl Zeiss make and Model No. EVO18) and TEM (FEI make and model no. TECNAI T20S TWIN).

3. Results and discussions

3.1. XRD analysis

The diffraction pattern, Figure 1, shows that the sample $Zn_{0.9}Mn_{0.05}Fe_{0.05}O$ is polycrystalline with wurtzite (hexagonal) structure having preferred orientations along (100), (002) and (101) planes and showing the growth of ZnO crystallites along different directions [16]. No other Phases except that for ZnO is detected, indicating that the doping of Mn and Fe does not alter the Wurtzite (hexagonal) structure of ZnO. The peaks intensity is sharp and narrow, confirming that the sample is of good quality with perfect crystallinity and fine grain size. The lattice parameters "a" and "c" are calculated for plane (100) and (002) of the sample. The values of lattice parameters are a=b= 3.2525 Å and c = 5.2071 Å. The values are slightly higher than those of reported for ZnO polycrystalline powder (a = b=3.249 Å, c = 5.206 Å) in reference [17].
The slightly increasing value of lattice parameters is due to the smaller ionic radii of Zn^{2+} (0.74Å) ions than Mn^{2+} (0.80Å) ions and comparable to Fe^{2+} ions [18] and secondly, the oxygen vacancy associated with nanoparticles [19]. The advanced structural parameter ratio c/a=1.60 is resulting due to hexagonal close packing. The standard value of the c/a ratio for the Wurtzite structure is about 1.63. It is known that when the bonding character is more ionic, the c/a ratio moves to the ideal values [20]. There is a good concurrence between obtained values and the standard one which indicates that Mn^{2+} and Fe^{2+} ions are able to substitute the Zn ions in the ZnO lattice without any effect on overall crystal structure. The obtained volume (V) = 47.7Å^3 of unit cell in $Zn_{0.9}Mn_{0.05}Fe_{0.05}O$ is more than that of pure ZnO, due to increase of the lattice parameters "a" and "c" [21, 22]. The Atomic packing fraction (APF) = 0.755 was calculated using the XRD data. The APF of particles is found to be nearly 75.5% in hexagonal structure, which is slightly more than the reported value 74% for ZnO [22]. The bond length (L) =

1.996 Å for $Zn_{0.9}Mn_{0.05}Fe_{0.05}O$ is calculated. It is known that ionic radius of O^{2-} is 1.38Å and ionic radius of Zn^{2+} is 0.74Å consequently, the length of the Zn-O bond is 2.12Å [18, 23, 24]. The small value of bond length indicates the presence of the structural defects or oxygen vacancies in our sample. The importance of the broadening of peaks indicates grain refinement along with the large strain coupled with the powder.

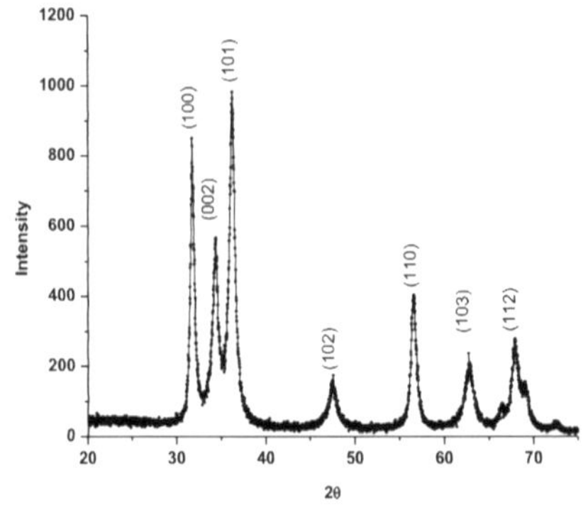

Figure 1. X-ray diffraction (XRD) of the sample $Zn_{0.9}Mn_{0.05}Fe_{0.05}O$.

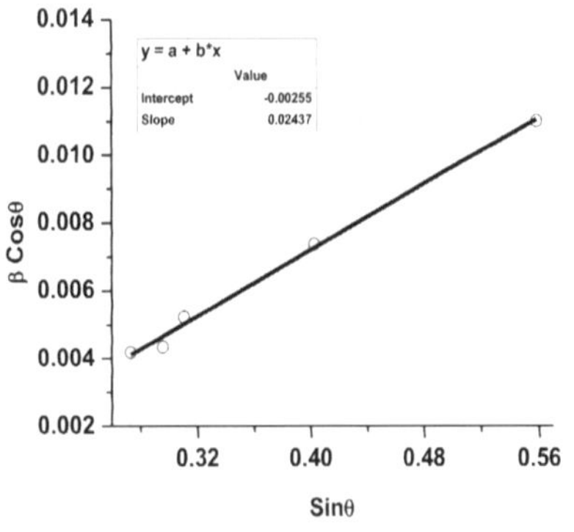

Figure 2. Williamson- Hall plot for the sample $Zn_{0.9}Mn_{0.05}Fe_{0.05}O$.

The instrument- corrected broadening (β_{hkl}) corresponding to the diffraction peak of $Zn_{0.9}Mn_{0.05}Fe_{0.05}O$ was estimated using the relation:

$$\beta^2_{hkl} = \beta^2_{hkl\,measured} - \beta^2_{instrumental},$$

The particle size (t_{DS}) in the samples is estimated using Debye Scherer's (DS) formula as in equation (1) [25-27].

$$t_{DS} = \frac{k\lambda}{\beta_{hkl}\cos\theta} \quad (1)$$

Where, t_{DS} is the particle size, k is the Scherer's constant and is taken equal to 0.9, λ (wavelength of $Cu_{K\alpha}$) is 0.15418 nm and β_{hkl} is full width at half-maximum (fwhm) of X-ray diffraction peaks in radians and θ is angle of Bragg's reflection.

Crystallite sizes along different crystallographic planes are presented in Table 1. The values of particle size lie in the range from 13 to 33nm indicating the nano-crystalline nature of the sample.

Table 1. Particle size determined by DS formula and W-H plot, and lattice strain in $Zn_{0.9}Mn_{0.05}Fe_{0.05}O$ sample

Sample	**Particle size (t_{DS}) in nanometre**						
	(100)	**(002)**	**(101)**	**(102)**	**(110)**	**(103)**	**(112)**
Gauss fitted fwhm (rad)	.0045	.0047	.0057	.0082	.0057	.0098	.0134
$Zn_{0.9}Mn_{0.05}Fe_{0.05}O$	33	32	27	19	29	17	13

Due to massively defect clusters of the grain edge and imperfection free of grain interior, the nano-crystalline material usually grieves from structural strain [28, 29].

The average crystallite size and strain can be calculated by Williamson and Hall [W-H] plot [29], as shown in Figure 2, with different X-ray peaks taken into consideration. Depending on different θ positions the size and strain broadening analysis has been done using [W- H] relation as in equation (2):

$$\beta_{hkl}\cos\theta = \frac{0.9\lambda}{t_{WH}} + 2\varepsilon\sin\theta \tag{2}$$

Where, t_{WH} is the crystalline size and ε is lattice strain. Graph between β_{hkl} cosθ and sinθ is plotted and then a linear fit to the points is extrapolated to give the crystallite size and Lattice strain [26, 31]. From W-H plot particle size (t_{WH}) = 54 nm and strain (ε) = 0.0122 were observed, Strain might be due to the lattice shrinkage. The particle size obtained from DS formula closely fitted with those obtained by the W-H plot for the sample under tensile strain. The dislocation density (ρ), represents the length of dislocation lines per unit volume, and the amount of defects [32,33], here varies randomly between 9×10^{14} lines/m^2 and 63×10^{14} lines/m^2 for $Zn_{0.9}Mn_{0.05}Fe_{0.05}O$ sample along each plane [Table 2]. The orientation parameter $\gamma_{[hkl]}$ for the $Zn_{0.9}Mn_{0.05}Fe_{0.05}O$ sample is obtained from the relative heights of all the peaks present in the pattern Table 2.

Table 2. Dislocation density (ρ), Orientation Parameter (γ[hkl]), Morphology index (MI) for $Zn_{0.9}Mn_{0.05}Fe_{0.05}O$

Sample $Zn_{0.9}Mn_{0.05}Fe_{0.05}O$	**Planes** **(100)**	**(002)**	**(101)**	**(102)**	**(110)**	**(103)**	**(112)**
Dislocation Density (ρ)	$9X10^{14}$	$10X10^{14}$	$14X10^{14}$	$28X10^{14}$	$12X10^{14}$	$35X10^{14}$	$63X10^{14}$
Orientation Parameter ($\gamma_{[hkl]}$)	0.24	0.16	0.28	0.05	0.12	0.07	0.08
Morphology Index (MI)	0.75	0.74	0.70	0.62	0.70	0.58	0.5

The values of $\gamma_{[hkl]}$ varies from 0.05 to 0.24 indicating random orientation. Larger orientation appears along the plane (101) in the sample [34, 35]. A XRD Morphology Index (MI) is derived from fwhm of XRD data. MI relates the fwhm of two peaks to its particle morphology. Generally, for highest fwhm peak, the MI is 0.5 because the MI is derived from the single peak. MI is obtained using equation (3).

$$\mathrm{MI} = \left[\frac{[\mathrm{fwhm}]_H}{[\mathrm{fwhm}]_H + [\mathrm{fwhm}]_P}\right] \tag{3}$$

Where, $[fwhm]_H$ is highest fwhm value obtained from peaks and $[fwhm]_P$ is value of particular peak's fwhm. The values of MI lie in the range of 0.50 to 0.75 and the details are presented in Table 2. It is correlated with the particle size (ranges from 13 to 33nm). It is observed that MI has linear relationship with particle size [36]. The result is shown in Figure 3.

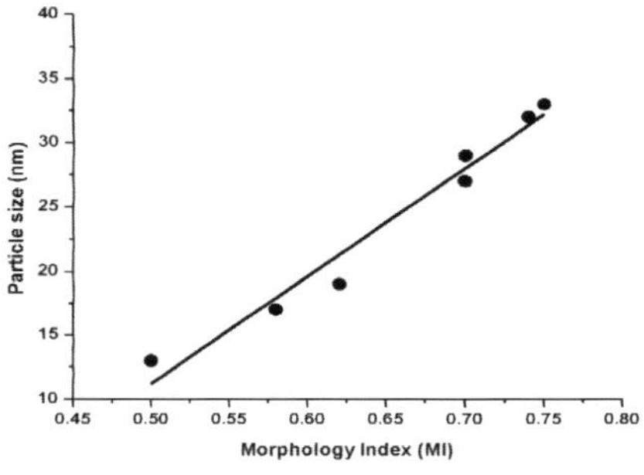

Figure 3. Morphology index versus Particle size.

Figure 4. SEM image of the $Zn_{0.9}Mn_{0.05}Fe_{0.05}O$ nanoparticles.

4. Morphological measurements

Scanning electron microscopy is used here to study the surface morphology of the $Zn_{0.9}Mn_{0.05}Fe_{0.05}O$ sample. Figure 4 SEM measurements shows the formation of cotton balls like structure of irregular shape and simultaneously agglomeration of particles seems to have occurred on the surface. SEM measurements shows the irregular shape of the nanoparticles and particle size lies in the nano range. TEM provides further insight into the morphologies and structural details of sample. Figure 5, shows the TEM image of the $Zn_{0.9}Mn_{0.05}Fe_{0.05}O$ nanoparticle. The observed size of $Zn_{0.9}Mn_{0.05}Fe_{0.05}O$ nanoparticles by TEM micrograph exhibits the similar trend with the size calculated by Debye-Scherer formula. TEM images also show the irregular shape of the nanoparticles and particle size lies in the range of 10-35nm. Similar to SEM image, the agglomeration of the nanoparticles has been observed from TEM also. Figure 6, displays the electron diffraction patterns of the sample which shows that the $Zn_{0.9}Mn_{0.05}Fe_{0.05}O$ particles nature are crystalline. Seven fringe pattern corresponding to planes (100), (002), (101), (102), (110), (103), and (112) are consistent with the peaks observed in the XRD pattern [37]. Figure 7, shows the HRTEM image of the $Zn_{0.9}Mn_{0.05}Fe_{0.05}O$ nanoparticle. HRTEM pattern indicates that the $Zn_{0.9}Mn_{0.05}Fe_{0.05}O$ has standard hexagonal structure and grow along the c-axis [38].

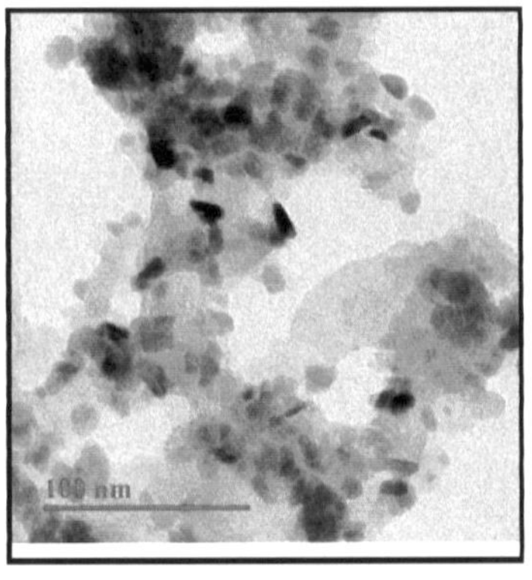

Figure 5. TEM image of the $Zn_{0.9}Mn_{0.05}Fe_{0.05}O$.

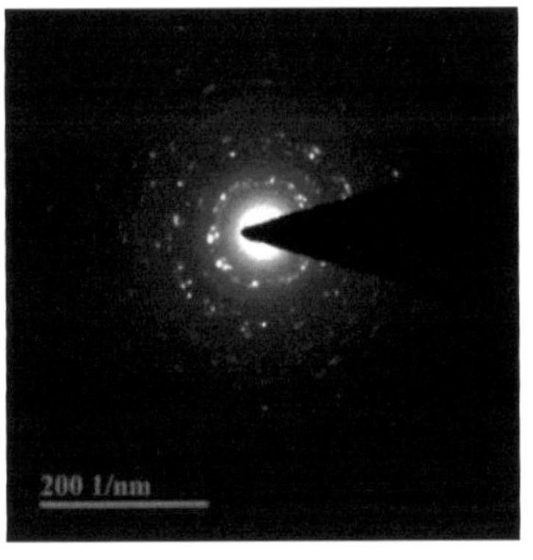

Figure 6. SAED image of $Zn_{0.9}Mn_{0.05}Fe_{0.05}O$.

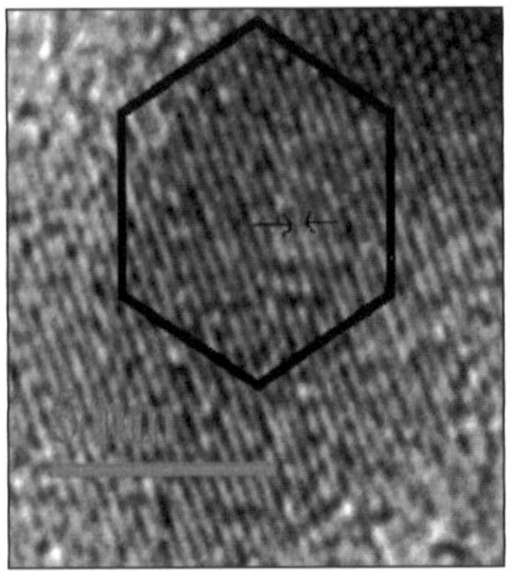

Figure 7. HRTEM pattern of $Zn_{0.9}Mn_{0.05}Fe_{0.05}O$.

5. Conclusions

The Sample $Zn_{0.9}Mn_{0.05}Fe_{0.05}O$ was synthesized successfully by Sol- gel wet chemical precipitation route. The crystal structure and morphological study of $Zn_{0.9}Mn_{0.05}Fe_{0.05}O$ sample were investigated at length. The $Zn_{0.9}Mn_{0.05}Fe_{0.05}O$ has wurtzite (hexagonal) crystal structure. The SEM image shows the clustering of particles occurred on the surface. TEM images show the irregular shape of the nanoparticles and particle size lies in the range of 10-35nm. HRTEM pattern indicates that the $Zn_{0.9}Mn_{0.05}Fe_{0.05}O$ has standard hexagonal structure and grow along the c-axis

6. References

[1] Gyu-Chul Yi, Wang C and Park W 2005 *Semicond. Sci. Technolog* **20** S22.

[2] Zhang Q, Yu K, Bai Q W, Wang Q, Xu F, Zhu Z, Dai N and Sun Y 2007 *Materials Lett.* **61**, 3890-3892.

[3] Nomura K, Ohta H, Ueda K, Kamiya T, Hirano M and Hosono H 2003 *Science* **300** (5623) 1269-1272 .

[4] Yuzhen L, Guo L, Huibin X, Ding L, Chunlei Y, Jiannong W, Weiken G, Shine Y and Ziyu W 2006 *J. Appl. Phys.* **99** 114302-114304.

[5] Ushio Y, Miyayama M and Yanagida H 1994 *Sensor and Actuat. B* **17** 221-226.

[6] Ohtomo A, Tamura K, Kawasaki M, Makino T, Segawa Y, Tang Z K, Wong G K L, Matsumoto Y and Koinuma H 2000 *Appl. Phys. Lett.* **77** 2204-2206.

[7] Pearton S J, Heo W H, Ivill M, Norton D P and Steiner T 2004 *Semicond. Sci. Technol.***19**, R59-R74.

[8] Dietl T, Ohno H, Matsukura F, Cibert J and Ferrand D 2000 *Science* **287**, 1019-1022.
[9] Liu C, Yun F and Morkoc H 2005 *J. of Mater. Sci.: Mater. in Electro.* **I6**, 555.
[10] Seta P D C, Yunita P E, Malik A B and Darminto 2017, *IOP Conf. Series: Materials Science and Engineering,* **196**, 01.
[11] Wang L and Muhammed M 1999 *J. Mater. Chem.*, **9**, 2871-2878.
[12] Han TY J, Worsley M A, Bauman T F and Satcher J H 2011 *J. Mater. Chem.* **21**, 330.
[13] Kumaran S M and Gopal Krishnan R 2012 *J. Sol-Gel Sci. Technol.* **62**, 193-200.
[14] Brinker C J and Scherer G W 1990 *"Sol–Gel Science Academic press,"* Boston
[15] Zarzycki J 1997 *J. Sol-Gel Sci. Technol.* **8**, 17-22.
[16] Cao L, Jiang J and Zhu L 2013 *Mater. Lett.* **100**, 201.
[17] Vaseem M, Umar A and Hann Y B 2010 *American Scientific Publishers, CA, USA* (ISBN 1-58883-170-1)
[18] Chang Y, Wang P, Sun Q, Wang Y and Long Y 2011 *J. of Nanomat.* 367156.
[19] Sundaresan A, Bhargavi R, Rangarajan N, Siddesh U and Rao C N R 2006 *Phys. Rev B* **74**, 161306 (R).
[20] Schulz H and Thiemann K H 1979 *Solid State Commun.* **32**, 783.
[21] Rusu D I, Rusu G G and Luca D 2011 *Acta Physica Polonica A* **6**, 119.
[22] Morkoc H and Ozgur U, *"Zinc Oxide: Fundamentals, Materials and Device Technology,"* Wiley-VCH Verlag GmbH & Co. ISBN: 978-3-527-40813-9
[23] Wang X S, Wu Z C and Webb J F 2003 *Appl. Phys.* **A77**, 561.
[24] Pearton S J, Norton D P, lp K, Heo Y W and Steiner T 2004 *J. Vac. Sci. Technol. B* **22**, 171.
[25] Barret C S and Massalski T B 1980 *"Structure of metals,"* Pergamon press, Oxford.
[26] Tripathi A, Misra K P and Shukla R K 2014 *J. of Lumin.* **149**, 361.
[27] Patterson A L 1939 *Phys. Rev. Online Arch. (Prola)* **56**, 978.
[28] Staab TEM, Krause- Rehberg R and Kieback B 1999 *J. Mater. Sci.* **34**, 3833.
[29] Ungar T, Tichy G, Gubicza J and Hellmi R J 2005 *Powder. Diffrac.* **20**, 366-375.
[30] Williamson G K and Hall W H 1953 *Acta. Metall.***1**, 22 -31.
[31] Srivastava A, Shukla R K and Misra K P 2011 *Cryst. Res.Technol.* **146**, 949.
[32] Khan Z R, Khan M S, Zulf Equar M and Khan M S 2011 *Mater. Sci. Appl.* **2**, 340-345.
[33] Ghosh A and Choudhary R N P 2009 *J. Appl. Phys.* 124906.
[34] Srivastava A, Kumar N and Khare S 2014 *Opto- Electron. Rev.* **22**, 68.
[35] Srivastava A, Kumar N, Misra K P and Khare S 2014 *Mater.Sci. in Semicon. Process.* **26**, 259.
[36] Prabhu Y T, Rao K V, Kumar V S and Kumari B S 2013 *Adv. Nanoparticles* **2**(01), 45
[37] Patil G E, Kajale D D, Gaikwad V B and Jain G H 2012 *Int. J. Nano Lett.* **2**, 17
[38] Aneesh P M, Vanaja K A, Jayaraj M K 2007 *Proc. of SPIE*, vol. **6639**, 66390J,

Acknowledgments
The authors are grateful to Management of Manipal University Jaipur for financial grant in the project and thankful to Materials research centre, MNIT, Jaipur and U.S.I.C, University of Rajasthan for SEM and TEM Measurements.

IOP Conf. Series: Materials Science and Engineering **348** (2018) 012005 doi:10.1088/1757-899X/348/1/012005

Nanotoxicity prediction using computational modelling - review and future directions

Bhavna Saini[1], Sumit Srivastava[2]

1 Manipal University Jaipur, Rajasthan, India
2 Manipal University Jaipur, Rajasthan, India.
E-mail: bhavna.saini@jaipur.manipal.edu

Abstract: Nanomaterials has stimulated various outlooks for future in a number of industries and scientific ventures. A number of applications such as cosmetics, medicines, and electronics are employing nanomaterials due to their various compelling properties. The unending growth of nanomaterials usage in our daily life has escalated the health and environmental risks. Early nanotoxicity recognition is a big challenge. Various researches are going on in the field of nanotoxicity, which comprised of several problems such as inadequacy of proper datasets, lack of appropriate rules and characterization of nanomaterials. Computational modelling would be beneficial asset for nanomaterials researchers because it can foresee the toxicity, rest on previous experimental data. In this study, we have reviewed sufficient work demonstrating a proper pathway to proceed with QSAR analysis of Nanomaterials for toxicity modelling. The paper aims at providing comprehensive insight of Nano QSAR, various theories, tools and approaches used, along with an outline for future research directions to work on.

1. Introduction

The intensely microscopic size of nanomaterials helps the cells to penetrate more easily, as well as models different chemical, physical, biological properties. These peculiarities of nanoparticles draw in the various industries to employ these particles in various applications such as cosmetics, pharmaceuticals, electronics etc. Nanomaterials can also be engineered for particular functionalities such as improving strength, catalyst nature, electrical and thermal conductivity etc. As the properties of nanomaterials are not always well characterized, they call for risk assessment of possible exposures arising during their manufacture and use.

Nanotoxicology is a branch of toxicology, which analyses the toxicity of nanomaterials. The study directs that to what extent the toxic effects of nanoparticles can impair the human health as well as environment. It is of vital importance to have knowledge of subsequent toxic effects of nanomaterials before utilizing it in applications. Every nanoparticle does not bring about the detrimental health effects, although there are various aspects on which the toxicity of nanoparticles depends such as size, configuration, surface functionality, particle ageing etc. [1][2].

The composition and size of nanomaterials plays a definite role in cellular response. In [3], the author has justified that the same material can generate different Intracellular responses and potential mechanism of toxicity depending on the exposed cell type. The compositions of nanomaterials get varied according to the products and brands. Table 1 shows some of the most commonly used nanomaterials along with their compositions, used in different items. This data has been collected from the Household Products Database of the National Library of Medicine which is based on the Consumer Product Information Database (NLM). The data of the table shows the average compositions used, but this does not prove helpful in defining the threshold values of these nanoparticles for toxicity check. This is due to the variation in the amount of the particles utilized in products. As mentioned in the table, different materials are used in different amounts for different products. The toxic effects of these materials come into existence after a regular use for a long time. The author in [4] formed a list of distinct cosmetics manufactured in different countries. It shows that composition again gets vary with the variation in manufacturing countries. The standards followed by various countries manufacturers and brands are different, due to which it is difficult to set a fixed usage value for these materials.

Table 1. Nanomaterials composition used in household products

Nanoparticle	Usage	Composition
Titanium Dioxide	Sunscreen (SPF 15)	1.0 – 5.0
Titanium Dioxide	Paint	5.0 – 30.0
Zinc Oxide	Sunscreen (SPF 30, 36)	Concentration up to 25% (Safe)
Zinc Oxide	Personal Care (Cream)	10.0 – 20.0
Silicon Dioxide	Sunblock lotion or cream SPF (30,45,55)	1.0 – 5.0
Silicon Dioxide	Personal Care (In form of paste or gel)	10.0 – 30.0
Silicon Dioxide	Personal Care (In form of liquid or solid)	0.1 - 1

Experimental data are critical to accelerate the insight of various properties of nanoparticles as well as for predictive analysis. The data, metadata and outcome of various experiments are present in different formats. Depicting such mix up data, requires a standard approach for compilation and communication. For most of the nano domain experts having mere informatics assistance, it will be really troublesome to programmatically execute it. The broad diversification and categorization of NM makes the toxicity interpretation more complicated. A minor change in the nanostructure properties, indicates the variation in their biological response, which again escalates the efforts to figure out the cause. For this, it is required to again go through each NM. The only feasible method to predict toxicity without experiment is to use their structural properties. QSAR (Quantitative Structure Activity Relationship) is a computational approach which endeavour to predict the biological functioning of a compound on the basis of their physiochemical properties and theoretical molecular descriptors. The next section discusses the physicochemical properties and descriptors which were studied and evaluated previously for toxicity prediction.

2. Properties affecting nanotoxicity

2.1. Physicochemical properties

Nanomaterials toxic effects are brought about by their various properties. For building up the toxicity prediction model, the first and foremost phase is to recognize the properties of nanoparticles which accelerate the toxic nature. The OECD group working on Nanomaterials had provided an extensive list of significant physicochemical properties, appropriate for toxicological studies [5].

Along with these basic properties, QSAR needs quantifier data to prevent incorrect interpretation. Table 1 index the elementary properties of nanoparticles (also known as physicochemical descriptors) and their quantifiers, which were analysed earlier for toxicity prediction. In [6][7][8][9] particle size was taken as base parameter and quantifiers as particle size and size distribution to find out toxicity level of TiO2 and Nanosilver. The study predicts that smaller size nanoparticles are more prone to toxicity as compared to larger size nanoparticles. Here the size distribution data gives better prediction results as compared to average size value. Therefore all the possible quantifiers need to be checked out while proceeding with QSAR approach. Table 2 displays the details of the analysis done in past, based on various properties such as shape, crystal structure, surface charge, surface functionality and aggregate phase. Other than these, there exist few more physicochemical properties such as purity, chemical composition, solubility, hydrophobicity, porosity, dustiness, production process and many more on which the toxic nature of a compound depends. [7][10]. More experimental data as well as research efforts are required to explore QSAR modelling based on the physicochemical properties of nanomaterials.

Table 2. Physicochemical properties affecting toxicity

Nanoparticles Properties	*Quantification*	*Examples*	*Observations*
Size [6][7][8][9]	Particle size Size Distribution	Tio2, Nano-silver	Smaller size exhibit high toxicity
Shape [8][11][12]	Spherical, Rectangle, Long, Short, Rod Like	Carbon-Nanotubes (CNT) , Silver NPs	Long CNT signify more toxicity as compared to short CNT
Crystal Structure [2][13]	Tetragonal, Orthorhombic, Monoclinic	TiO2, Nano Silica	Toxicity may alter contingent upon structure
Surface Charge [14][15]	Positive, Negative, Weakly Negative, neutral	Silicon NPs, Silica NPs	Negatively charge NPs are more toxic as compared to neutral and positively charged NP
Surface Functionality [16][17][18][19][20]	Coating, Customization	Ag-NPs	NP Surface can be functionalized deliberately which can alter the toxic nature.
Aggregate Phase [21][22][23][24]	Solid, dry, liquid, suspension	NPs impulse to contact each other and form chunks	Collective nature affects various critical properties, so it needs to be considered.

2.2. Nano descriptors

For designing a proper computational model, only physicochemical properties are not sufficient. As nano features differ a lot, there is a need to look upon the nano descriptors (NP Descriptors). NP Descriptors can be defined as peculiar characteristic or consolidation of various characteristics of a materials. Table 3 exhibit some of the NP Descriptors used in the past [25][26][27] took microscopic images as NP descriptors. These microscopic images are the blend of various characteristics such as surface scope, aspect ratio, shape, roughness, pixels, RGB codes and chemical ordering. On the basis of these features' values, the level of harmfulness can be predicted.

Table 3. NP descriptors affecting toxicity

NP-Descriptors	*Characteristics*
Microscopic Images [26]	Surface scope, aspect ratio, shape, roughness, chemical ordering
Microscopic Images [27]	Pixels, RGB codes, Grayscale
Biological Surface Absorption Index [28]	Polarity, lone pair electron, hydrogen bond acceptor and donor
Spectra Category [25]	Infrared, Ultraviolet
Procedure Followed [29]	Development, Combination and Transformation methods used

Similarly, other NP descriptors were defined as the combination of various characteristics of nanomaterials. Not only properties, but the procedures followed during development, combination and transformation of compound used as the initial data to explore the toxicity level.
These descriptors data may not be in appropriate form to apply on computational models, so it needs to be refined. One of the main issue is to translate these theoretical descriptors into computer language for computation. The complication and heterogeneity of NM make it difficult to transform initial data into standard formats. Some standard methods and formats are required to refine these raw data into computer accessible format. Next section discusses the previous work done on collection of nanomaterials data from various resources as well as on processing it into a common format.

3. Nanodata collection and format conversion

In order to pull together with this diversified nanomaterials study, some formats and standards are required for extensive communication of the data. Several associations are working in this directions. This section gives the overview of the past work done in this area.
A framework was introduced i.e. ISA-TAB Nano (Investigation – Study – Assay – Material Tab delimited) [3]. ISA-Tab Nano is an extension of ISA-Tab, which is a general purpose framework introduced for proper collection and communication of experimental data.
ISA-Tab Nano utilizes a sequence of technologies which collect and transmit complex nanomaterial data in a standard way. It accepts the data into four files namely investigation file, study file, assay file and materials file which are further divided into specific fields or columns. This data manually accumulated from the literature on predefined templates and various business rules are also applied. Additionally, a python program is present to convert resulting ISA-Tab nano files to text files delimited using tab. This output file assist the progress of computational analysis and database compliance. The issues associated with it includes the need of revision of various recorded data such as reaction rate, chemical compositions as well as template design. Another problem is that this scheme does not suit well for all type of experimental data and vivo toxicology study. The data collected is not good enough for further analysis. It needs to be redefine, to get utilized in predictive modelling.
In [30] the author had footnoted the requirement for more quality efforts in collecting the suitable and quality nano data as well as to develop prediction models for nanotoxicity check. The authors included few NanoQsar examples applied in past, to explain the prediction model development process. However, this analysis is not sufficient to provide a pathway in this area.
As fitting high quality dataset is one of the key prerequisite in the area of nanomaterials research, lemin xiao et al. tried to solve this dilemma by developing an information extraction system [31]. This extraction system performs carried out ontology-based entity extraction and rule-based attribute extraction on publications relevant to nanotoxicity. Further the extracted data is stored into a table accordance with relations among the data. As this system is based on ontology, it is not adequate to discover new entity instances. Machine learning methods can prove beneficial in determining new instances and inflate the dictionary.

4. Computational techniques

The past segment talks about the different approaches to gather and arrange the test information. To investigate the accessible information, different computational or information mining systems are required to smooth the assessment procedure. In [4][32][33], the authors have reviewed the QSAR approaches applied in the past. The following segment examines the most generally utilized strategies for the prediction of nanotoxicity.
Regression: Regression analysis is widely used in prediction and forecasting. It explains the variation occurs in value of dependent variables with respect to independent variable. The most widely used linear regression algorithms include multiple linear regression (MLR), Partial least square regression (PLR) and Principle component regression (PCR) [34][35][36][37]. The PLS regression is used when there is inter correlation between certain descriptors. These strategies demonstrates advantageous when different reactions should have been displayed simultaneously. The foremost advantage of linear regression methods is their transparency to provide relevant significance information of

physicochemical descriptors. . Once in a while these models turn out to be ineffectively performed as a result of the nearness of information factors which are other than the yield. Consequently to expel these different measurement lessening strategies, for example, PCA (Principle Component Analysis) is utilized. Different past examinations had utilized PCA to enhance the result of their QSAR demonstrate.
Decision tree: A decision tree is an efficient approach used in classification and regression. In QSAR, it can be used to predict toxicity on the basis of numerical and categorical data of biological activities [38][39]. The significant capabilities of decision tree which contribute the most in toxicity prediction includes automatic selection of input variables, removal of insignificant descriptors and to find worthwhile out of small, noisy and large datasets.
Support vector machine: Various issues such as collinear descriptors, nonlinear relations, small and large datasets as well as over fitted models are handled by SVM. SVM proves to be an efficient approach for linear modelling which can manage both classification and regression problem [40][41]. The capability of SVM to give high precision and speculation making it an approach of intrigue.
Artificial neural network: ANN is another approach utilized for nonlinear data relationship and large datasets. ANN approach has certain disadvantages such as choosing the ideal complexity, trouble of overfitting, high generalization sensitivity to variation in parameters and network topology and last but not the least, trouble in foreseeing the result [42][43].

5. QSAR (Quantitative structure activity relationship)

Nastassja A. et al. had analysed the distinct informatics mechanisms that have been adapted in patent mining, nanomaterial device characterization, nanomedicine, and environmental risk assessment [44]. They single out the various natural language processing (NLP)-based tools such as NanoPort, NanoMapper, TechPerceptor, Nanodevice Analyser, Nanotoxicity Searcher, NEIMiner etc. Some of which are no longer available to use. The authors have endorsed for sharing NLP-related tools through online repositories to broaden participation in nanoinformatics.
A new model GA-SVM, which works with the combination of two algorithms i.e. genetic algorithm and support vector machine was proposed [45]. The main aim of this paper is to classify the nanomaterials toxicity based on size, surface area and charge, chemical composition structure, reactivity and partitioning characteristics. The results obtained from experiments were satisfactory, but this model is applicable only when training data is available. While working with new set of experimental descriptors, it does not prove fruitful.
Xiong Liu et al. proposed a prediction cube approach for nanotoxicity modelling [47]. Instead of using whole datasets (as done in traditional modelling), this data cube make use of subset modelling. It performs data exploration and predictive analytics which can help researchers in
Slicing, dicing and drilling through nanotoxicity data and measure the quality of different subsets for building prediction models. Additionally, it stores prediction models and results in different cube cells. The inadequacy of proper experimental data and complications in acquiring that data, led to QSAR.
QSAR (Quantitative structure-activity relationship) is defined as a mathematical model, which endeavour to generate a computable algorithm using various biological properties, compound structure and physico chemical features. As toxicity can be allied with such measurable properties, several QSARs models were proposed.
The initial efforts to put QSAR in use for nanotoxicity computing was done by [46]. They checked for different toxicity sources using QSAR analysis, but due to unavailability of sufficient experimental data it did not turn out convincing.
In [49], the author has come up with a QSAR model. The author has presented that various machine learning methods can generate promising predictive models for nanoparticles' biological reactions. Three different datasets were taken as input that includes iron oxide nanoparticles, gold nanoparticles and other nanoparticles adorned with different molecules and are having different properties such as compositions, coating, concentration etc. Here Bayesian neural network along with linear and nonlinear machine learning techniques were adopted.
Further various potential QSAR models were developed. In [29], the authors had proposed an effective statistical QSAR Model to anticipate the nanotoxicity. Moreover various classification based QSAR models were formed [48][50].

Natalia et al. have imparted a worthwhile review on QSAR approaches [4]. The paper focuses on inorganic NPs and provides an informative chemometrics data summary. As the lack of availability of proper datasets is one of the major issue in nanoinformatics, this review had tried to guide through the available and previously used datasets in QSAR, QSPR and chemometric studies. Various related publications were investigated and the summary is explained with the use of distinct fields such as no. of NPs, descriptors used, learning process and origin of datasets are some of the fields which are presented in the paper.
Oksel et al in [33][52] provides the details about core requirements for QSAR model. They have reviewed the characterization of nanomaterials, in-silico tools for QSAR model and summary of previous studies in this area. This paper can prove helpful while working on nanotoxicity computational modelling.

6. Conclusion

Nano QSAR needs a lots of efforts at every step, from pre-processing to model development. A biggest challenge is the lack of proper and sufficient experimental data. The major reason behind this inadequacy of data, is the complication in standardizing the toxicity testing methods and characterization of properties. The characterization of nanoparticles plays important role not only in identification of relationship between nanostructure and biological activities, but also in hastening the computation of engineered NM. There is a need for standard computation models, which collect experimental data and convert it to standard formats, to process it further for prediction model. The paper tries to be outright in its scope by reviewing considerable work in the field of NanoQSAR. We have analysed the previous work to bridge the gap between two research areas that are nanotoxicity and computational modelling, by providing the base as well as fusion of these two areas. Various techniques applied in past for experimental data collection, conversion to standard formats, QSAR and NanoQSAR are discussed in detail. To carry out comprehensive utilization of research data from bioscience groups, interoperable and efficient techniques are required which also help in building up an open data communing culture. We encourage the use of data mining techniques in the field of nanotoxicity, to programmatically execute the experimental data and predict the toxicity of nanomaterials before their manufacture and use.

7. References

[1] Sohaebuddin, S. K., Thevenot, P. T., Baker, D., Eaton, J. W., & Tang, L. (2010). Nanomaterial cytotoxicity is composition, size, and cell type dependent. Particle and fibre toxicology, 7(1), 1.
[2] Jiang, J., Oberdörster, G., Elder, A., Gelein, R., Mercer, P., & Biswas, P. (2008). Does nanoparticle activity depend upon size and crystal phase?. Nanotoxicology, 2(1), 33-42.
[3] Thomas, D. G., Gaheen, S., Harper, S. L., Fritts, M., Klaessig, F., Hahn-Dantona, E., ... & Klemm, J. D. (2013). ISA-TAB-Nano: a specification for sharing nanomaterial research data in spreadsheet-based format. BMC biotechnology, 13(1), 1.
[4] Sizochenko, N., & Leszczynski, J. (2016). Review of Current and Emerging Approaches for Quantitative Nanostructure-Activity Relationship Modeling: The Case of Inorganic Nanoparticles. Journal of Nanotoxicology and Nanomedicine (JNN), 1(1), 1-16.
[5] OECD. (2010). Guidance Manual for the Testing of Manufactured Nanomaterials: OECD's Sponsorship Programme. In. Paris: Organization for Economic Co-operation and Development.
[6] Park, M. V., Neigh, A. M., Vermeulen, J. P., de la Fonteyne, L. J., Verharen, H. W., Briedé, J. J., ... & de Jong, W. H. (2011). The effect of particle size on the cytotoxicity, inflammation, developmental toxicity and genotoxicity of silver nanoparticles. Biomaterials, 32(36), 9810-9817.
[7] Pettitt, M. E., & Lead, J. R. (2013). Minimum physicochemical characterisation requirements for nanomaterial regulation. Environment international, 52, 41-50.
[8] Powers, K. W., Palazuelos, M., Moudgil, B. M., & Roberts, S. M. (2007). Characterization of the size, shape, and state of dispersion of nanoparticles for toxicological studies. Nanotoxicology, 1(1), 42-51.

[9] Karlsson, H. L., Gustafsson, J., Cronholm, P., & Möller, L. (2009). Size-dependent toxicity of metal oxide particles a comparison between nano-and micrometer size. Toxicology letters, 188,112-118.
[10] Powers, K. W., Carpinone, P. L., & Siebein, K. N. (2012). Characterization of nanomaterials for toxicological studies. Nanotoxicity: Methods and Protocols, 13-32.
[11] Monteiro-Riviere, N. A., & Tran, C. L. (Eds.). (2007). Nanotoxicology: characterization, dosing and health effects. CRC Press.
[12] Gratton, S. E., Ropp, P. A., Pohlhaus, P. D., Luft, J. C., Madden, V. J., Napier, M. E., & DeSimone, J. M.(2008). The effect of particle design on cellular internalization pathways. Proceedings of the National Academy of Sciences, 105, 11613-11618.
[13] Napierska, D., Thomassen, L. C., Lison, D., Martens, J. A., & Hoet, P. H. (2010). The nanosilica hazard: another variable entity. Particle and fibre toxicology, 7(1), 39.
[14] Park, Y. H., Bae, H. C., Jang, Y., Jeong, S. H., Lee, H. N., Ryu, W. I., ... & Jeong, J. (2013). Effect of the size and surface charge of silica nanoparticles on cutaneous toxicity. Molecular & Cellular Toxicology, 9(1), 67-74.
[15] Bhattacharjee, S., de Haan, L. H., Evers, N. M., Jiang, X., Marcelis, A. T., Zuilhof, H. & Alink, G. M. (2010). Role of surface charge and oxidative stress in cytotoxicity of organic monolayer-coated silicon nanoparticles towards macrophage NR8383 cells. Particle and fibre toxicology, 7(1), 25.
[16] Caballero-Díaz, E., Pfeiffer, C., Kastl, L., Rivera-Gil, P., Simonet, B., Valcárcel, M., ... & Parak, W. J. (2013). The toxicity of silver nanoparticles depends on their uptake by cells and thus on their surface chemistry. Particle & Particle Systems Characterization, 30(12), 1079-1085.
[17] Nguyen, K. C., Seligy, V. L., Massarsky, A., Moon, T. W., Rippstein, P., Tan, J., & Tayabali, A. F. (2013). Comparison of toxicity of uncoated and coated silver nanoparticles. In Journal of Physics: Conference Series (Vol. 429, No. 1, p. 012025). IOP Publishing.
[18] Silva, T. U. (2011). An evaluation of coating material dependent toxicity of silver nanoparticles
[19] Yang, X., Gondikas, A. P., Marinakos, S. M., Auffan, M., Liu, J., Hsu-Kim, H., & Meyer, J. N. (2011). Mechanism of silver nanoparticle toxicity is dependent on dissolved silver and surface coating in Caenorhabditis elegans. Environmental science & technology, 46(2), 1119-1127.
[20] Zhao, C. M., & Wang, W. X. (2012). Importance of surface coatings and soluble silver in silver nanoparticles toxicity to Daphnia magna. Nanotoxicology, 6(4), 361-370.
[21] Dhawan, A., & Sharma, V. (2010). Toxicity assessment of nanomaterials: methods and challenges. Analytical and bioanalytical chemistry, 398(2), 589-605.
[22] Jiang, J., Oberdörster, G., & Biswas, P. (2009). Characterization of size, surface charge, and agglomeration state of nanoparticle dispersions for toxicological studies. Journal of Nanoparticle Research, 11(1), 77-89.
[23] Von der Kammer, F., Ferguson, P. L., Holden, P. A., Masion, A., Rogers, K. R., Klaine, S. J., ... & Unrine, J. M. (2012). Analysis of engineered nanomaterials in complex matrices (environment and biota): general considerations and conceptual case studies. Environmental Toxicology and Chemistry, 31(1), 32-49.
[24] Boverhof, D. R., & David, R. M. (2010). Nanomaterial characterization: considerations and needs for hazard assessment and safety evaluation. Analytical and bioanalytical chemistry, 396(3), 953-961.
[25] Burello, E., & Worth, A. P. (2011). QSAR modeling of nanomaterials. Wiley Interdisciplinary Reviews: Nanomedicine and Nanobiotechnology, 3(3), 298-306.
[26] Glotzer, S. C., & Solomon, M. J. (2007). Anisotropy of building blocks and their assembly into complex structures. Nature materials, 6(8), 557-562.
[27] Puzyn, T., Leszczynska, D., & Leszczynski, J. (2009). Toward the development of "nano-QSARs": advances and challenges. Small, 5(22), 2494-2509.
[28] Xia, X. R., Monteiro-Riviere, N. A., & Riviere, J. E. (2010). An index for characterization of nanomaterials in biological systems. Nature nanotechnology, 5(9), 671-675.
[29] Epa, V. C., Burden, F. R., Tassa, C., Weissleder, R., Shaw, S., & Winkler, D. A. (2012). Modeling biological activities of nanoparticles. Nano letters, 12(11), 5808-5812.

[30] Richarz, A. N., Madden, J. C., Robinson, R. L. M., Lubiński, Ł., Mokshina, E., Urbaszek, P., & Cronin, M. T. (2015). Development of computational models for the prediction of the toxicity of nanomaterials. Perspectives in Science, 3(1), 27-29.

[31] Xiao, L., Tang, K., Liu, X., Yang, H., Chen, Z., & Xu, R. (2013, December). Information extraction from nanotoxicity related publications. In Bioinformatics and Biomedicine (BIBM), 2013 IEEE International Conference on (pp. 25-30). IEEE.

[32] YOSEFU, N. O. (2015). COMPUTATIONAL MODELLING FOR PREDICTION OF NANOMATERIAL TOXICITY (Doctoral dissertation, Makerere University).

[33] Oksel, C., Ma, C. Y., Liu, J. J., Wilkins, T., & Wang, X. Z. (2017). Literature Review of (Q) SAR Modelling of Nanomaterial Toxicity. In Modelling the Toxicity of Nanoparticles (pp. 103-142). Springer International Publishing.

[34] Eriksson, L., & Johansson, E. (1996). Multivariate design and modeling in QSAR. Chemometrics and intelligent laboratory systems, 34, 1-19.

[35] Gu, C., Goodarzi, M., Yang, X., Bian, Y., Sun, C., & Jiang, X. (2012). Predictive insight into the relationship between AhR binding property and toxicity of polybrominated diphenyl ethers by PLS-derived QSAR. Toxicology letters, 208, 269-274.

[36] Shahlaei, M. (2013). Descriptor selection methods in quantitative structure?activity relationship studies: a review study. Chemical reviews, 113, 8093-8103.

[37] Yee, L. C., & Wei, Y. C. (2012). Current Modeling Methods Used in QSAR/QSPR. assessment, 10, 11.

[38] Wang, X. Z., & Ma, C. Y. (2009). Morphological population balance model in principal component space. AIChE Journal, 55, 2370-2381.

[39] Bengio, Y., Delalleau, O., & Simard, C. (2010). Decision trees do not generalize to new variations. Computational Intelligence, 26, 449-467.

[40] Mei, H., Zhou, Y., Liang, G., & Li, Z. (2005). Support vector machine applied in QSAR modelling. Chinese Science Bulletin, 50, 2291-2296.

[41] Darnag, R., Minaoui, B., & Fakir, M. (2012). QSAR models for prediction study of HIV protease inhibitors Using Support Vector Machines, Neural Networks and Multiple Linear Regression. Arabian Journal of Chemistry.

[42] Ventura, C., Latino, D. A., & Martins, F. (2013). Comparison of Multiple Linear Regressions and Neural Networks based QSAR models for the design of new antitubercular compounds. European journal of medicinal chemistry, 70, 831-845.

[43] Sussillo, D., & Barak, O. (2013). Opening the black box: low-dimensional dynamics in high-dimensional recurrent neural networks. Neural computation, 25, 626-649

[44] Lewinski, N. A., & McInnes, B. T. (2015). Using natural language processing techniques to inform research on nanotechnology. Beilstein journal of nanotechnology, 6(1), 1439-1449.

[45] Oladele, O. T. (2014). Nanomaterials Characterization Using Hybrid Genetic Algorithm Based Support Vector Machines.

[46] Liu, J., & Hopfinger, A. J. (2008). Identification of possible sources of nanotoxicity from carbon nanotubes inserted into membrane bilayers using membrane interaction quantitative structure–activity relationship analysis. Chemical research in toxicology, 21(2), 459-466.

[47] Liu, R., Rallo, R., Weissleder, R., Tassa, C., Shaw, S., & Cohen, Y. (2013). Nano-SAR development for bioactivity of nanoparticles with considerations of decision boundaries. Small, 9(9-10), 1842-1852.

[48] Liu, R., Zhang, H. Y., Ji, Z. X., Rallo, R., Xia, T., Chang, C. H., & Cohen, Y. (2013). Development of structure–activity relationship for metal oxide nanoparticles. Nanoscale, 5(12), 5644-5653.

[49] Winkler, D. A. (2016). Recent advances, and unresolved issues, in the application of computational modelling to the prediction of the biological effects of nanomaterials. Toxicology and applied pharmacology, 299, 96-100.

[50] Singh, K. P., & Gupta, S. (2014). Nano-QSAR modeling for predicting biological activity of diverse nanomaterials. RSC Advances, 4(26), 13215-13230.

[51] Oksel, C., Ma, C. Y., Liu, J. J., Wilkins, T., & Wang, X. Z. (2015). (Q) SAR modelling of nanomaterial toxicity: A critical review. Particuology, 21, 1-19.
[52] Oksel, C., Ma, C. Y., & Wang, X. Z. (2015). Current situation on the availability of nanostructure–biological activity data. SAR and QSAR in Environmental Research, 26(2), 79-94.
[53] Buzea, C., Pacheco, I. I., & Robbie, K. (2007). Nanomaterials and nanoparticles: Sources and toxicity. Biointerphases, 2(4), MR17-MR71.
[54] Kleandrova, V. V., Luan, F., González-Díaz, H., Ruso, J. M., Melo, A., Speck-Planche, A., & Cordeiro, M. N. D. (2014). Computational ecotoxicology: Simultaneous prediction of ecotoxic effects of nanoparticles under different experimental conditions. Environment international, 73, 288-294.
[55] Liu, X., Tang, K., Xiao, L., Song, M., & Xu, R. (2014, November). Nanotoxicity modeling in multidimentional cube. In Bioinformatics and Biomedicine (BIBM), 2014 IEEE International Conference on (pp. 25-28). IEEE.
[56] NLM: The National Library of Medicine: https://householdproducts.nlm.nih.gov/index.htm
[57] Oksel, C., Winkler, D. A., Ma, C. Y., Wilkins, T., & Wang, X. Z. (2016). Accurate and interpretable nanoSAR models from genetic programming-based decision tree construction approaches. Nanotoxicology, 10(7), 1001-1012.
[58] Puzyn, T., Rasulev, B., Gajewicz, A., Hu, X., Dasari, T. P., Michalkova, A., ... & Leszczynski, J. (2011). Using nano-QSAR to predict the cytotoxicity of metal oxide nanoparticles. Nature nanotechnology, 6(3), 175-178.
[59] .SLiu, X., Tang, K., Harper, S., Harper, B., Steevens, J. A., & Xu, R. (2012, October). Predictive modeling of nanomaterial biological effects. In Bioinformatics and Biomedicine Workshops (BIBMW), 2012 IEEE International Conference on (pp. 859-863). IEEE.
[60] Winkler, D. A., Burden, F. R., Yan, B., Weissleder, R., Tassa, C., Shaw, S., & Epa, V. C. (2014). Modelling and predicting the biological effects of nanomaterials. SAR and QSAR in Environmental Research, 25(2), 161-172.
[61] Winkler, D. A., Mombelli, E., Pietroiusti, A., Tran, L., Worth, A., Fadeel, B., & McCall, M. J. (2013). Applying quantitative structure–activity relationship approaches to nanotoxicology: current status and future potential. Toxicology, 313(1), 15-23.
[62] Ying, J., Zhang, T., & Tang, M. (2015). Metal oxide nanomaterial QNAR models: available structural descriptors and understanding of toxicity mechanisms. Nanomaterials, 5(4), 1620-1637.

IOP Conf. Series: Materials Science and Engineering **348** (2018) 012006 doi:10.1088/1757-899X/348/1/012006

Evaluation of ionic liquid treated sisal (*agave sisalana*) fiber as sorbent in biodiesel spill

E Costa[1], A Silva[2] and S Mattedi[1]

1 Graduate Program in Chemical Engineering, Federal University of Bahia, Salvador, Bahia, Brazil
2 Materials Science and Engineering Department, Federal University of Bahia, Salvador, Bahia, Brazil
E-mail: silvana@ufba.br

Abstract. The global economic development continually demands the use of energy resources, among which various types of oils are widely used. Despite their undeniable economic importance, environmental accidents with these occur frequently. Thus, the search for efficient and low-cost mitigating measures is recurrent. In this context, techniques that use natural adsorbents, such as vegetable fibers, have been studied, since they combine efficiency, selectivity, low cost and sustainability. Studies have been carried out using various types of fibers, natural or chemically treated. The interest in treating the fiber lies in the fact that, changing the chemical structure of the fiber, its oil sorption capacity is increased. Due to the offered advantages, an alternative and promising type of surface treatment using ionic liquids was performed, to the detriment of traditional treatments. Thus, the technical feasibility of the use of sisal fiber treated with ionic liquid for adsorption of biodiesel was studied.

1. Introduction

The use of different types of oils as energy sources began many years ago and these are extremely important for world economic development. Despite the predominant use of fossil fuels and their derivatives, the search and encouragement for the use of cleaner energy sources, such as biodiesel, is increasing. The unstoppable demand for any type of fuel, however, leads increasing problems due to its production, transportation and use. It is undeniable that the burning of biodiesel produces the least amount of pollution, when comparing to petroleum subproducts. However, spills of this biofuel in the marine and / or terrestrial environment are as dangerous as those of other oils, not only causing great damage to the local fauna and flora, but also having a negative impact on the regional economy, leisure and tourism. Therefore, technologies that can reverse this situation in a Sustainable and low cost way[1-4] are extremely necessary, which demonstrates the importance of studies in this field.
Numerous measures to reverse this type of accident are currently known. Mechanical, physical, chemical and biological processes are already employed by oil production and refineries, which aim to recover, remove or degrade the spilled oil. Among these processes, the most used ones are natural dispersion, containment and cloning, on-site burning, adsorption and use of detergents and dispersants. The use of adsorbent materials in this context, allows the oil to be removed in an efficient and selective manner, allied to a low cost of operation, and even sustainable, depending on the material used in the recovery.[5]
The adsorption phenomenon can be defined as a process in which a substance called adsorbate (solid, liquid or gaseous) binds, physically or chemically, to the surface of another solid substance known as adsorbent. Adsorption has been under increasing interest of researchers, since this technique has been

shown to be quite efficient in the treatment of organic and heavy metal effluents[6,7]. The adsorbent materials used can be of synthetic or natural origin and are classified as organic minerals, organic synthetic and natural sorbents, depending on the type of material used[8,9]. However, it is the use of natural materials that have been the focus of countless researches lately, because besides presenting high adsorption capacity, they are biodegradable, obtained through renewable sources, of low cost and widely available in the market. The fibers, which are of vegetable origin, are sustainable sorbent materials, which can be justified due to their biodegradability and the possibility of reuse. Moreover, these materials are highly efficient in oil adsorption, due to their great porosity and hydrophobic nature3. Brazilian areas for agricultural activity are quite extensive and the large production of these consumer goods provides the country with a wide possibility of using natural fibers as adsorbents. In this context, sisal (agave sisalana) fiber, for example, can be widely used in the sorption of oil spilled in the environment, due to its high production in Brazil and the enormous amount of waste generated. Some researchers have developed studies on the use of natural fibers as oil adsorbents and have proved the technical efficiency of using these materials to remedy the problem in question. The vegetable fibers are composed of three main components, being them cellulose, hemicellulose and lignin. The first two, due to the large amount of hydroxyl molecules (-OH) in its structure, have a hydrophilic character. The lignin is responsible for the degree of hydrophobicity of the fiber. Thus, in order to improve the sorption capacity of vegetable fibers by oils, surface treatments are used to modify the chemical structure of these adsorbent materials, making them more hydrophobic[10-13]. In the treatment of vegetable fibers for oil sorption, the traditional techniques are acetylation and mercerization. Despite their efficiency, the chemical agents used in the process present disadvantages such as toxicity, aggressiveness on the fibers and high cost. In this scenario, the ionic liquids appear as a new option of chemical treatment of fibers, due to low volatility, thermal stability, besides the possibility of adjustment of properties, such as polarity and miscibility. Researches have shown the efficiency of the ionic liquids in the dissolution and modification of the cellulose [14-17], however few works were performed with pre-treatment for natural adsorbents.

2. Methodology

2.1 Materials

The ionic liquid (IL) 2-hydroxy ethylammonium acetate ([2HEA] [Ac]) was synthesized for this work, and its structure was confirmed by NMR 1H, and FTIR, which was synthesized for this work. The biodiesel used was provided by Process Separation Laboratory (UFBA) [18]. It was synthesized from soybean oil by a methanolic synthesis route18. Sisal (agave sisalana) fibers were supplied as sisal ropes by APAEB a cooperative of small rural producers from Valente, Bahia, Brazil.

2.2 Fiber treatment

The treatment of the sisal fibers was done in a thermostatic bath for 2 hours at 353.15K, with the immersion of a certain amount of fiber [14] in rope form in the solution of ionic liquid and distilled water (0.75 g IL/g solution). After the treatment time, the fibers were washed with distilled water until wash water had a neutral pH. Finally, the treated fibers were placed in a greenhouse at 333.15K for 24 hours. The surface morphology was studied through SEM images before and after treatment.

2.3 Kinetics and adsorption equilibrium

To evaluate biodiesel sorption by the treated fiber, 0.5g of the fiber in a solution of biodiesel/water was placed in a thermostatic finite bath at 303.15K for 1 hour .The mass of the fiber was weighed before and after the sorption process, which was calculated by the following mathematical expression:

$$S = \frac{(m_f - m_i)}{m_i} \quad (1)$$

For adsorption kinetics, the fiber was kept in contact with the oil at different times (5 to 90min), while in equilibrium experiments the amount of oil in the solution (1 to 10ml) was varied. These tests were done with the treated and untreated fibers and the sorption was calculated in all cases through

Equation 1. These tests have the objective of evaluating the mechanism of the adsorption process that occurs between the fiber and oil, besides measuring the maximum amount of oil that can be adsorbed.

3. Results and discussion

3.1 Surface modification

SEM images of the treated and untreated fibers are shown in Figures 1-2. It could be observed that the treatment was effective and removed wax and impurities from fiber surface thus altering the surface area and exposing the internal fibrillar structure enhancing oil sorption by the fiber.

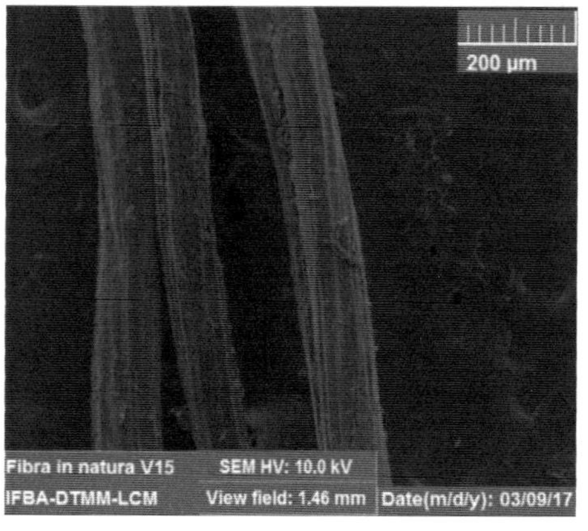

Figure 1. SEM micrograph of untreated sisal fiber.

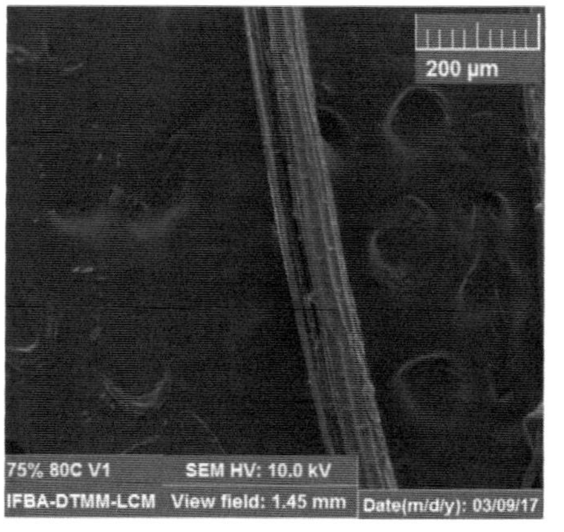

Figure 2. SEM micrograph of treated sisal fiber.

3.2 Sorption kinetics

The adsorption kinetics tests, using biodiesel, were performed for treated and untreated sisal fibers. Based on Figure 3, it can be inferred that the use of vegetable fibers to adsorb oils lighter and less viscous than petroleum, in this case biodiesel, is also efficient. In addition, an increase in the sorption of this oil was obtained by using the treated fiber, demonstrating the effectiveness of the treatment with the ionic liquid.

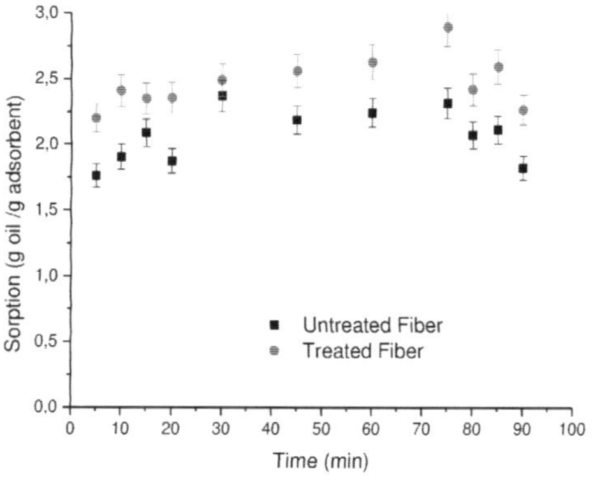

Figure 3. Sorption Kinetics – Biodiesel.

The modeling of the experimental kinetic data was done through three models: pseudo-first order, pseudo-second order and intraparticular diffusion[19-22]. These models were chosen because they are the most used in the literature and have good correlations in general. The three models were applied to the experimental data, and the best fit occurred with the pseudo-second order model. The correlation factors presented a very good value, besides small errors for the calculated parameters, as can be observed in figures 4 and 5.

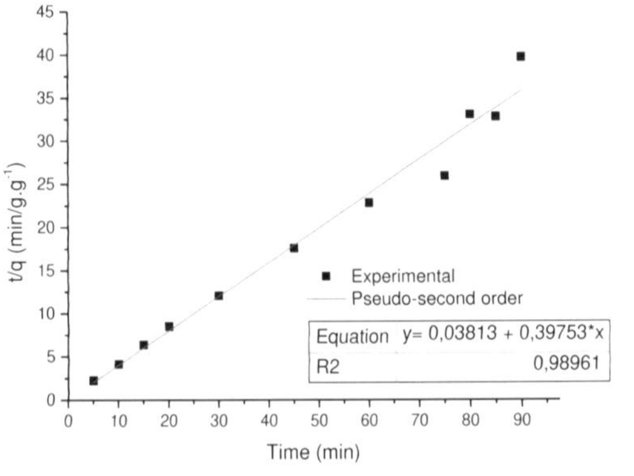

Figure 4. Kinetic model– treated fiber/ biodiesel.

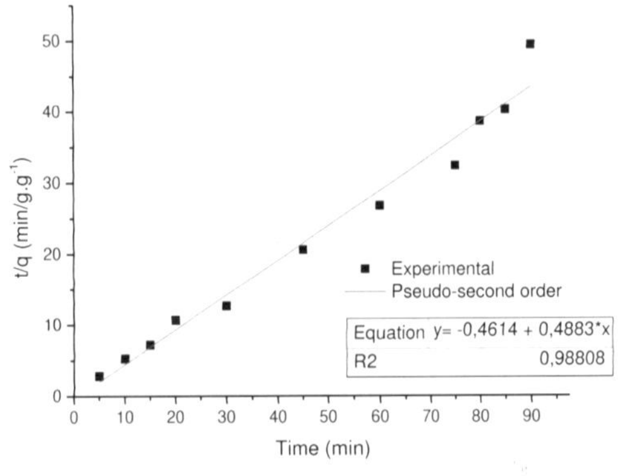

Figure 5. Kinetic model – untreated fiber/ biodiesel.

This result indicates that the adsorption occurred between sisal fiber and biodiesel was a chemical adsorption, and the reaction itself is the limiting stage of the process.

3.3 Equilibrium of sorption

In the equilibrium tests, the amount of biodiesel in a final solution of 100ml was varied. From these results, it is possible to identify the maximum degree of sorption by the fiber, important information when defining the necessary amount of adsorbent to suck the spilled oil. In the figures 6 and 7, the equilibrium data for the treated and untreated fibers were presented, as well as the modeling of these data through four models: Freundlich, Langmuir, Sips and Toth. The first two models were chosen because they are the most traditional, while the last two are adaptations of these first, correcting some of their failures[23,24].

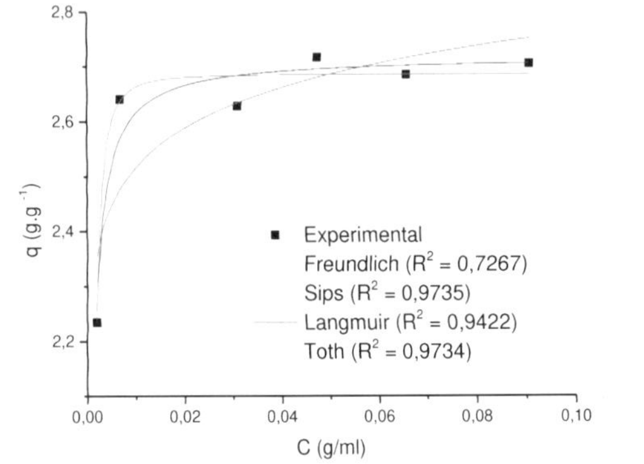

Figure 6. Equilibrium of sorption – treated fiber.

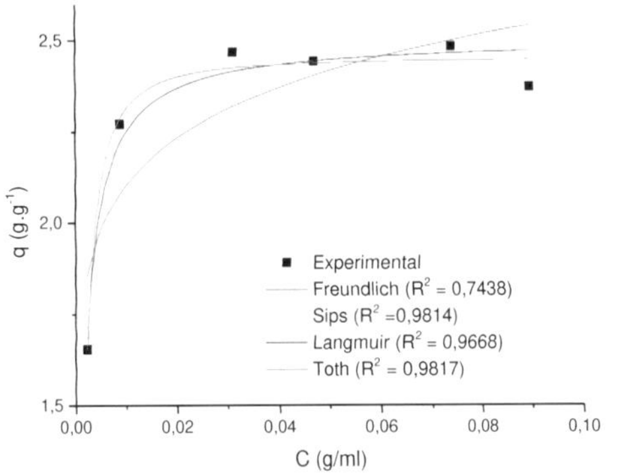

Figure 7. Equilibrium of sorption – untreated fiber.

Among the models used, those of Sips and Toth were the ones that best adjusted the experimental points, whose curves are practically superposed. While the Toth model minimizes the deviations between the experimental and theoretical points present in the Langmuir model, the Sips model is a combination of the Freundlich and Langmuir models. Depending on the adsorbate concentration, the Sips model falls into one model or the other. In the present case, in which there is a high concentration, the Sips model relies on the Langmuir model. Because of this, it is that the Sips and Toth curves are practically overlapping and the Langmuir model also presented a good value for the correlation factor, but lower than for Sips and Toth. These results suggest that the adsorption, in this case, happens in monolayers, besides indicating a heterogenous adsorptive surface, from the energetic point of view.

4. Conclusion

Considering the results obtained, the use of new solvents, such as ionic liquids, less toxic and less aggressive, proved effective in the treatment of these fibers, providing an increase in oil sorption. Because it is a very small area still studied, the use of other ionic liquids, combined with new treatment conditions, could produce even better results. In addition, it can be concluded that the use of vegetable fibers for adsorption of oils, which is already studied mainly for petroleum, is also effective in the case of oils that are lighter and less viscous. This signals that studies with these sorts of adsorbents should continue in order to improve this technique and make it as viable as using synthetic adsorbents. It was also possible to conclude from this work that the adsorption occurred between sisal fiber and biodiesel was of the chemical type, which reduces the possibility of desorption, since the interactions between the components in question are strong. Finally, the fact that adsorption occurs in monolayers is favorable, as it also decreases the level of desorption, and the presence of active sites with different energies allows the effective interaction of the adsorbent with different types of adsorbates.

5. References

[1]Ali I, Asim M, Khan T A 2012 *Journal of Environmental Management* Low cost adsorbents for the removal of organic pollutants from wastewater 113 170

[2] Said A E A, Ludwick A G, Aglan, H A 2009 *Bioresource Technology* Usefulness of raw bagasse for oil absorption: A comparison of raw and acylated bagasse and their components 100 2219

[3] Brandao P C, Souza T C, Ferreira C A, Hori C E, Romanielo L L 2010 *Journal of Hazardous Materials* Removal of petroleum hydrocarbons from aqueous solution using sugarcane bagasse as adsorbent 175 1106

[4] Paulauskienė T, Jucikė I, Juščenko N, Baziukė D 2014 *Water Air Soil Pollut* The use of natural sorbents for spilled crude oil and diesel cleanup from the water surface 225 1

[5] Sayed S A, Zayed A M 2006 *Desalination* Investigation of the effectiveness of some adsorbent materials in oil spill clean-ups 194 90

[6] Lakherwal D 2014 *International Journal of Environmental Research and Development* Adsorption of Heavy Metals: A Review 4 41

[7] Hegazi H A 2013 *Journal Housing and Building National Research Center* Removal of heavy metals from wastewater using agricultural and industrial wastes as adsorbents 9 276

[8] Sirotkina E E, Novoselova L Y 2005 *Chemistry for Sustainable Development* Materials for Adsorption Purification of Water from Petroleum and Oil Products 13 359

[9] Moyo F, Tandlich R, Wilhelmi B S, Balaz S 2014 *International Journal Environmental Research and Public Health* Sorption of Hydrophobic Organic Compounds on Natural Sorbents and Organoclays from Aqueous and Non-Aqueous Solutions: A Mini-Review 11 5020

[10] Albinante S R, Pacheco E B A, Visconte, L L Y 2013 *Quimica Nova* A review on chemical treatment of natural fiber for mixing with polyolefins In Portuguese 36 114

[11] Ciacco G T, Morgado D L, Frollini E, Possidoniob S, El Seoud O A 2010 *Journal Brazilian Chemical Society* Some Aspects of Acetylation of Untreated and Mercerized Sisal Cellulose 21 71

[12] Arsyad M, Wardana N G, Irawan Y S 2015 *Revista Materia* The morphology of coconut fiber surface under chemical treatment 20 169

[13] Hill C A S, Khalil H P S, Hale M D 1998 *Industrial Crops and Products* A study of the potential of acetylation to improve the properties of plant fibres 8 53

[14] Rocha E G A, Costa A C, Aznar M 2014 *Chemical Engineering Transactions* Use of Protic Ionic Liquids as Biomass Pretreatment for Lignocellulosic Ethanol Production 37 397
[15] Badgujar K C, Bhanage B M 2015 *Bioresource Technology* Factors governing dissolution process of lignocellulosic biomass in ionic liquid: Current status, overview and challenges 178 2
[16] Li J, Luo M, Zhao C J, Wang W, Zu Y, Fu Y 2013 *Bioresource Technology* Oil removal from water with yellow horn shell residues treated by ionic liquid 128 673
[17] Ohno H, Fulcaya Y 2009 *Chem. Lett.* Task especific ionic liquids for cellulose technology 38 2
[18] Simonelli G, Lima Lobato A K C, Santos L C L 2016 *International Journal of Advances in Science, Engineering and Technology* Development of an alternative biofuel from soybean oil 4 12
[19] Ferreira L P, Moreira A N, Delazare T, Oliveira, G E, Souza, F G 2012 *Macromolecular Symposia* Petroleum Absorbers Based on CNSL, Furfural and Lignin - The Effect of the Chemical Similarity on the Interactions among Petroleum and Bioresins 319 210
[20] Grance E G O, Souza F G, Varela A, Pereira E D, Oliveira G E, Rodrigues, C H M 2012 *Journal of Applied Polymer Science* New Petroleum Absorbers Based on Lignin-CNSL-Formol Magnetic Nanocomposites 126 304
[21] Santos E G, Alsina O L S, Silva, L H 2007 *Química Nova* Biomass performance in the adsorption of light hydrocarbons in aqueous effluents (in Portuguese*)* 30 327
[22] Wei Q F, Mather R R, Fotheringham A F, Yang R D 2003 *Marine Pollution Bulletin* Evaluation of nonwoven polypropylene oil sorbents in marine oil-spill recovery 46 780
[23] Bhagya N P, Prashanth P A, Raveendra R S, Sathyanarayani S, Ananda S, Nagabhushana B M, Nagabhushana H 2016 *Journal of Asian Ceramic Societies* Adsorption of hazardous cationic dye onto the combustion derived SrTiO3 nanoparticles: Kinetic and isotherm studies 4 68.
[24] Tian Y, Wu M, Liu R G, Wang D Q, Lin X B, Liu W L, Ma L, Li Y D, Huang Y 2011 *Journal of Hazardous Materials* Modified Native Cellulose Fibers – A Novel Efficient Adsorbent for Both Fluoride and Arsenic 185 93

Acknowledgments
The authors gratefully acknowledge the financial supports by FAPESB/SECTI-BA (Fundação de Amparo a Pesquisa do estado da Bahia) and CNPq/Brazil (Grant 306640/2016).

Photocatalytic water splitting over titania supported copper and nickel oxide in photoelectrochemical cell; optimization of photoconversion efficiency

Norani Muti Mohamed[1], Robabeh Bashiri[1*], Chong Fai Kait [2], Suriati Sufian[3]

1Centre of Innovative Nanostructures & Nanodevices (COINN), Universiti Teknologi PETRONAS, 32610 ,Seri Iskandar, Perak, Malaysia
2 Fundamental and Applied Sciences Department, Universiti Teknologi PETRONAS, 32610 ,Seri Iskandar, Perak, Malaysia
3Chemical Engineering Dep., Universiti Teknologi PETRONAS, 32610 ,Seri Iskandar, Perak, Malaysia
E-mail: robabeh.bashiri@utp.edu.my

Abstract. we investigated the influence of fluctuating the preparation variables of TiO_2 on the efficiency of photocatalytic water splitting in photoelectrochemical (PEC) cell. Hydrothermal associated sol–gel technique was applied to synthesis modified TiO_2 with nickel and copper oxide. The variation of water (mL), acid (mL) and total metal loading (%) were mathematically modelled using central composite design (CCD) from the response surface method (RSM) to explore the single and combined effects of parameters on the system performance. The experimental data were fitted using quadratic polynomial regression model from analysis of variance (ANOVA). The coefficient of determination value of 98% confirms the linear relationship between the experimental and predicted values. The amount of water had maximum effect on the photoconversion efficiency due to a direct effect on the crystalline and the number of defects on the surface of photocatalyst. The optimal parameter ratios with maximum photoconversion efficiency were 16 mL, 3 mL and 5 % for water, acid and total metal loading, respectively.

1. Introduction

Utilizing solar energy through photovoltaic technologies is one of the most attractive renewable energy sources to solve the worldwide energy issues. The early 1970s, Honda and Fujishima established solar-driven hydrogen production from water splitting over TiO_2 in a photoelectrochemical (PEC) cell. However, the remained challenge is the low efficiency of converting solar to chemical energy. Therefore, a semiconductor with low bandgap energy to absorb a wide region of the solar spectrum and negative conduction band is essential to enhance the photoconversion efficiency of water splitting [1]. Amongst various metal oxides, TiO_2 photocatalyst has received considerable attention as the key material in photoelectrochemical devices while its wide band gap energy (3.2 eV) and the relatively high recombination rates of photo-generated electron-holes restrain its different applications [2, 3]. Incorporation of dopant in the host to create a new energy state is the most feasible method to modify the physical and electronic properties of TiO_2 [4]. The 3d transition metals such as Cu, Fe, etc. are the well-known dopants to enhance the photocatalytic water splitting performance due to their large availability and low cost [5]. Riaz et al. [6] reported that the incorporation both Cu and Ni in the crystal structure of TiO_2 enhanced its photocatalytic degradation performance compared to single Cu or Ni doped and undoped TiO_2. The extensive research works reveal that the photocatalytic performance of TiO_2 can be altered by type of metal dopants and total metal loading, preparation techniques, experimental conditions, etc. [7]. The preparation parameters during sol-gel method play important role on the photocatalytic performance of TiO_2 as we reported the effect of three preparation parameters like water, acid, amount of Cu loading on the physical properties, photoelectrochemical behavior, and hydrogen production of Cu –Ni doped TiO_2 in a PEC/DSSC (dye sensitized solar cell) [8]. Herein, we investigated the effect of single and interaction between parameters (water, acid, and total metal loading) on the photoconversion efficiency in PEC cell using design of experiment (DOE).

2. Experimental section

Figure 1 shows the flowchart of optimization process through design of experiments. The outcome of experimental design was total of 20 experiments by varying the three preparation parameters. The experimental results were analyzed in which the relationship between photoconversion efficiency and parameters was modelled using ANOVA to find the greatest photocatalyst preparation condition for further optimization. Finally, this simulated work was compared with previous related works.

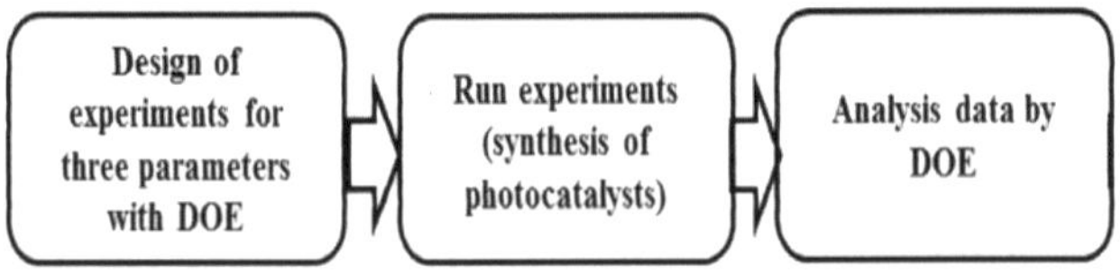

Figure 1. Flowchart of optimization process of preparation parameters.

Figure 2 shows the schematic diagram of all designed experiments in CCD using the following equation;

$$2^3\left(\text{factorial points}\right)+2? \left(\text{axial or star points}\right)+6\left(\text{centre points: six replications}\right)=20$$

The experimental errors was be estimated based on the number of independent variables, usually between 3 and 10 [9]

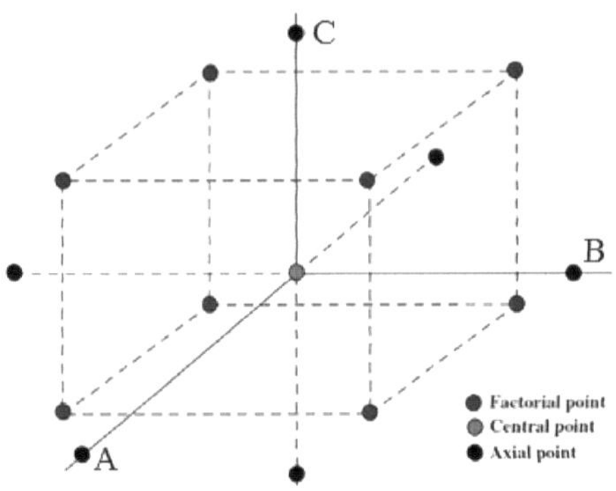

Figure 2. Schematic diagram of CCD as a function of A, B and C [10].

The experiments were designed using Design - Expert software version 8.0.7.1 (Stat-Ease, Inc.) and The coded and actual values of three investigated parameters in CCD are shown in Table 1.

Table 1. The real and coded levels of the independent process parameters in CCD

Factors	Factors Code	Coded levels of variables				
		-2(α)	-1	0	+1	+2 (+α)
Water:	A	16	32	48	64	80
Acid:	B	1	2	3	4	5
Total metal loading	C	3	5	9	12.5	15

IOP Conf. Series: Materials Science and Engineering **348** (2018) 012007 doi:10.1088/1757-899X/348/1/012007

The photocatalysts were synthesized via sol-gel associated hydrothermal method. Two solutions were prepared; (i) a mixture of Titanium tetraisopropoxide (TTIP), ethanol, glacial acetic, and $Ni(NO_3)_2 \cdot 6H_2O$ and $Cu(NO_3)_2.3H_2O$ and (ii), a mixture of water and ethanol. The solution (ii) was added dropwise to the solution (i) under vigorous string and the resulting hydroxide precipitates were directly transferred to autoclave for heat treatment process in an oven at a temperature of 180 °C for 12 h. Afterward, the autoclave was cooled down in room temperature and participates was centrifuged and rinsing with deionized water and ethanol. Finally, the products were dried at 105 °C over the night, calcined at 450 °C for 120 min, and labelled based on the run number from 1 to 20 [11-13]. In order to prepare a printable paste, the synthesized photocatalysts (Cu-Ni/ TiO_2) were mixed with water, acetic acid, ethanol, polyethylene glycol 20000, Eethylcellulose, and α –Terpineol and then printed on the fluorine doped tin oxide (FTO) glass substrate using the screen printing method. We reported the details of paste preparation in our previous works [1, 14].

2.1. Photoelectrochemical measurements

Figure 3 shows the experimental setup, including prepared thin film of Cu-Ni/TiO_2 as a working electrode, Pt rod as counter electrode, and filled Ag/AgCl with saturated KCl as a reference electrode. The electrodes were immersed in a mixture of 1M KOH and 10 % glycerol and they connected to a potentiostat (Autolab PGSTAT302N Metrohm) with three-electrode configurations. The experiments were conducted under a 150 W xenon lamp with intensity of 1 sun (100 mW/cm^2) and filtered by AM 1.5G filter to eliminate the UV (wavelength < 400 nm) region of the solar spectrum. The photoconversion efficiency for each cell was calculated from $\eta\ \% = j_p\ [(E^o_{rev} - |E_{app}|)\ /I_o$. The photocurrent density (j_p) was measured with Linear Sweep Voltammetry (LSV) procedure with applied voltage from 0.4 to 0.4 V vs. Ag/AgCl and the scan rate of 20 mV/s. Furthermore, E^o_{rev}= 1.23 V is standard potential of water and the absolute value of the applied potential ($|E_{app}|$) vs. reversible hydrogen electrode (RHE) obtained when the working electrode potential at the measured photocurrent under illumination was misused from open-circuit condition versus RHE [14].

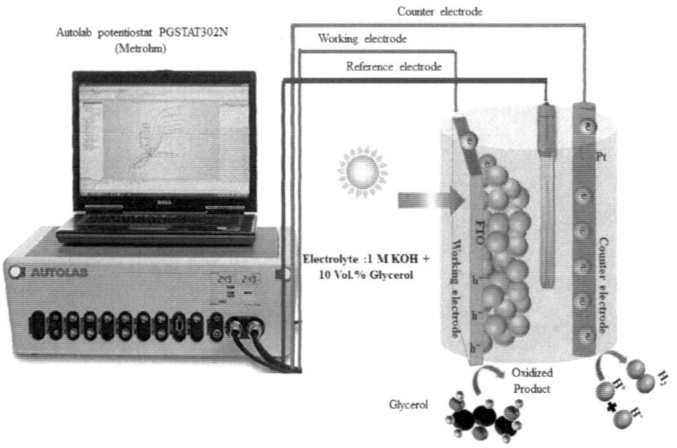

Figure 3. The photoelectrochemical study setup and electrode configuration.

3. **Result and discussions**

3.1. Statistical analysis of photoconversion efficiency

The model of photoconversion efficiency (η %) is represented in the following equation:

$$\eta\%= +1.02 - (1.01\times A) + (0.27\times B) - (0.2\times C) + (0.5\times A \times B) + (0.17\times A\times C) + (0.093\times B \times C) + (1.4 \times A^2) - (0.67\times B^2) + (0.05\times C^2)$$

The summary of ANOVA analysis, including P-value (significance probability value) and student-tests as suggested by Montgomery to identify the significance of coefficient for each term are tabulated in Table 2.

Table 2. Summary of ANOVA for photoconversion efficiency (η %)

Source	Sum of mean square	[a]DF	Mean Square	F Value	p-value (Prob > F)	Comments
Model	10.76	9	1.20	80.28	< 0.0001	Significant
Residual	0.15	10	0.015			-
[b]LOF	0.043	4	0.011	0.6	0.613	not significant
R^2=0.986, Adj-R^2=0.97, Adeq Precision = 41.55, and [c]C.V.% = 9.7						

[a]DF = degree of freedom, [b]LOF =lack of fit, and [d]C.V. = coefficient of variation

This model is significant at 95% confidence interval with the F-value of 80.28 and this profound amount of F -value has less than 0.01% chance to occur because of noise. Furthermore, the P-value model is less than 0.0001< 0.05, indicating an extremely significant confidence level of 99.9%. However, F-value of lack of fit (LOF) is 0.6 and non-significant owing to pure error and there is a 67.39% chance to occur due to noise, thus this non-significant lack of fit is good to fit the model. Moreover, there is considerable agreement between the experimental and predicted data due to the high coefficients of determination (R^2) at 0.986 [24]. The R-squared illustrates that parameters have 98.6% impact on the photoconversion efficiency and the remaining responses of 1.4% is related to residues. The model has adequate signal with adequate precision ratio of 41.55 because Adeq-precision measures the signal to noise ratio and a ratio greater than 4 is desirable. This model is reproducible due to C.V. (9.7%) smaller than 10% [25]. Figure 4 and Table 3 indicate that there is a linear correlation between the actual vs. predicted values of photoconversion efficiency, confirming the best fit and the reliability of the model.

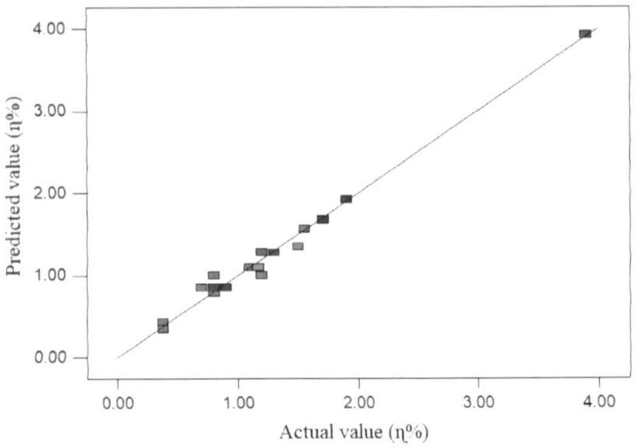

Figure 4. Correlation between actual and predicted values of photoconversion efficiency in CCD.

Table 3. Experimental and predicted values of photoconversion efficiency

(Photoconversion efficiency (η %))					
Run	**Actual value**	**Predicted value**	**Run**	**Actual value**	**Predicted value**
1	1.70	1.67	11	1.17	1.10
2	1.90	1.92	12	0.80	0.79
3	0.90	0.85	13	1.50	1.35
4	0.80	0.85	14	0.70	0.85
5	1.30	1.28	15	0.38	0.34
6	0.87	0.85	16	1.20	1.00

7	3.80	3.92	17	1.55	1.56
8	1.20	1.28	18	0.38	0.43
9	1.10	1.1	19	0.82	0.85
10	1.70	1.69	20	0.80	1.00

3.2. Overall evaluation based on model terms and 3D graphs

Table 4 presents the coefficient, P-value of each term, and interaction between them in the quadratic model. It is notable that the term with highest coefficient and smallest P-value has the maximum effect on the photoconversion efficiency in the PEC cell. In addition, the term with the positive coefficient has synergetic influence on the response, however; the term with negative coefficient has antagonistic effect on the response. Amongst the linear terms, water with highest navigate coefficient and the smallest P- value had maximum antagonistic effect on the photoconversion efficiency.

Table 4. P-value and coefficient of each parameter and interaction between them

Linear			Interaction terms			Quadratic		
Factor	Coefficient	p-value	Factor	Coefficient	p-value	Factor	Coefficient	p-value
A	-1.01	< 0.0001	AB	0.50	0.0429	A^2	1.47	< 0.0001
B	0.4	0.0021	AC	0.17	0.0108	B^2	-0.67	< 0.0001
C	-0.20	< 0.0001	BC	0.091	0.0264	C^2	0.054	< 0.0001

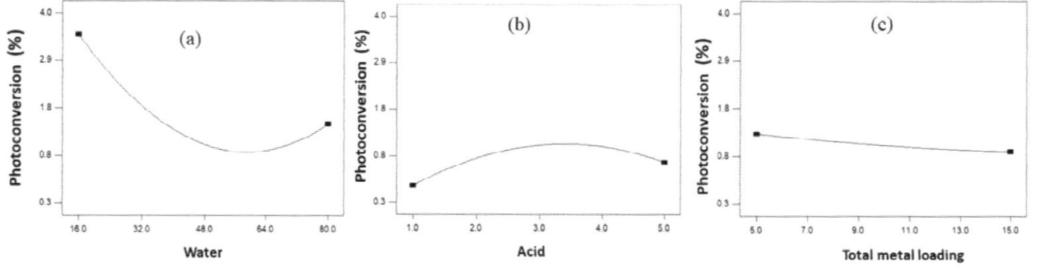

Figure 5. Single influence of each parameters (a) water, (b) acid, and (c) total metal loading on photoconversion efficiency.

Furthermore, Figure 5 (a) illustrates that water as the most important parameter had completely quadratic influence on the photoconversion efficiency with the greatest difference between the highest and lowest level of water around 1.5%. In addition, AB and A^2 terms had greater effects on the photoconversion efficiency compared to other interaction and quadratic terms. Moreover, a comparison between Figure 5 (b) and (c) shows the parameter B had more quadratic negative impact on the response with difference between the maximum and minimum level of 0.5% compared to parameter C. Figure 5 (c) shows that increasing the amount of total metal loading had negative impact on the photoconversion efficiency. Figures 6(a) depicts interaction between parameters A (water) and B (acid) molar ratio on the photoconversion efficiency at the centre point of C. It displays that the interaction between water and acid at a low level of water is more significant and this interaction had maximum positive impact on the photoconversion efficiency compared to the other interactions. Figure 6(b) and (c) reveal that the 3-D response surface plot and its respective contour plot for the interaction between A with C at the centre point of B and B with C at the centre point of A, respectively. Figure 6(b) depicts that the interaction between water and total metal loading was the second most important parameters and increasing amount of water as the lowest level of total metal loading had the negative impact on photoconversion efficiency of the PEC cell. In addition, Figure 6(c) demonstrates that the interaction B and C had the smallest impact on the photoconversion efficiency compared to A and B.

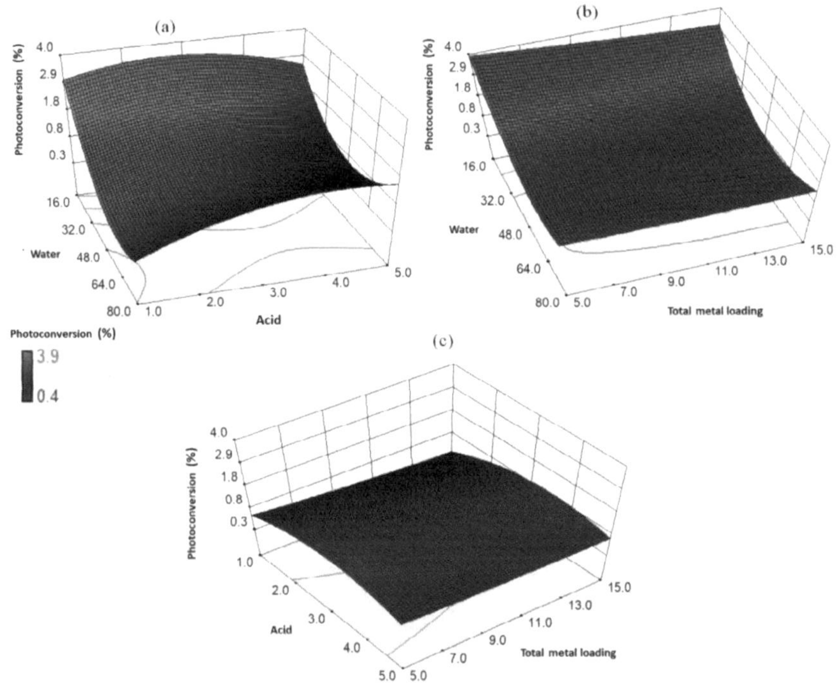

Figure 6. 3D response surface of interaction between parameters.

The negative effect of increasing the amount of water is corresponding to fast hydrolysis rate, formation aggregate particles, and less crystallinity of TiO_2 [2]. Moreover, adding acid as chelating agent has the synergetic influence on response because increasing the concentration of acid can moderate the activity of titania precursor, control the hydrolysis rate, form primary particles with higher surface area, and well crystallinity [15, 16]. On the other hand, adding Cu and Ni in the structure TiO_2 has some advantages like shifting absorbance of TiO_2 to the visible region, decreasing particle sizes and increasing the photocatyst surface are to absorb more active species on the surface of electrolyte [12]. However, finding an optimum level of total metal loading is essential because increasing total metal loading causes more charge transfer resistance, less electron lifetime and charge carrier density and decrease photoconversion efficiency [17]. The photoconversion efficiency results confirm that the prepared photocatalyst with preparation condition of run No.7 (water:16 mL, acid: 3 mL, and total metal loading of 5 mol%) had maximum photoconversion efficiency performance of 3.8%. The photoconversion efficiency of the prepared photoanode with the best performance was compared with some related works. The photoconversion efficiency of the PEC cell with an undoped TiO_2 photo-anode was calculated 0.4% and 0.13-2% in different electrolytes with applied voltage by Fujishima et al. [18] and Giordano et al. [19], respectively. Ghosh et al. reported that doping of TiO_2 with Al slightly increased the photoconversion efficiency from 0.4% to 1.3% [20]. Sun et al. [21] conclude that the presence of Sn on TiO_2 nanorod can change efficiency from 0.48% for undoped TiO_2 to 0.69% for Sn doped TiO_2 [21]. The difference of efficiency for water splitting strongly related to generation, separation, and transportation charge carriers in the surface of thin film [22]. The main reasons of such significant performance from run No.7 could be related to the synergetic effect of two different metals to trap electron–hole pairs. Since copper and nickel oxides act as electron and hole trapping centres, respectively, thus these phenomena can reduce recombination electron-hole pairs and improve electron transfer to the surface of photocatalyst [23].

4. Conclusion

The optimization of preparation variables (water, acid, and total metal loading) in sol-gel method was performed using central composite design from response surface method. The maximum photoconversion efficiency of 3.8% was calculated for prepared photoanode with water of 16 mL, acid of 3 mL, and total metal loading of 5 mol%. The quadratic models with R^2 value of 0.98 was considered the significant model to fit the data with better prediction responses. The linear and quadratic terms of water had more antagonistic impact on the response compared to B and C. The interaction AB and AC had higher significant influence on the photoconversion efficiency in compassion with AC and BC.

5. References

[1] Bashiri R, Mohamed NM, Fai Kait C, Sufian S, Khatani M. J Environ Chem Eng. 2017;**5**:3207-3214.
[2] Bashiri R, MutiMohamed N, Fai Kait C, Sufian S. Adv. Mat. Res.. 2014;**925**:396-400.
[3] Samsudin MFR, Sufian S, Mohamed NM, Bashiri R, Wolfe F, Ramli RM. Mater Lett. 2018;**211**:13-16.
[4] Hou X-G, Huang M-D, Wu X-L, Liu A-D. Chem Eng J. 2009;**146**:42-48.
[5] Gombac V, Sordelli L, Montini T, Delgado JJ, Adamski A, Adami G, et al. J Phys Chem A. 2009;**114**:3916-3925.
[6] Riaz N, Fai Kait C, Dutta BK, Man ZB, Khan MS, Nurlaela E. Chem Eng J. 2012;**185–186**:108-119.
[7] Feng H, Zhang M-H, Yu LE. Appl Catal, A. 2012;**413–414**:238-244.
[8] Bashiri R, Mohamed NM, Fai Kait C, Sufian S, Kakooei S, Khatani M, et al. Renew Energy. 2016;**99**:960-970.
[9] Singh KP, Gupta S, Singh AK, Sinha S. Chem Eng J. 2010;**165**:151-160.
[10] Cho I-H, Zoh K-D. Dyes Pigments. 2007;**75**:533-543.
[11] Bashiri R, Mohamed NM, Fai Kait C, Sufian S. Adv Mat Res. 2014;**925**:396-400.
[12] Bashiri R, Mohamed NM, Fai Kait C, Sufian S, Khatani M. Int J Hydrogen Energy. 2017;**42**:9553-9566.
[13] Bashiri R, Mohamed NM, Fai Kait C, Sufian S, khatani M, Hanaei H. Procedia Eng. 2016;**148**:151-157.
[14] Mohamed NM, Bashiri R, Fai Kait C, Sufian S, Kakooei S. Int J Hydrogen Energy. 2015;**40**:14031-14038.
[15] Liao DL, Liao BQ. J Photochem Photobiol, A. 2007;**187**:363-369.
[16] Canterino M, Di Somma I, Marotta R, Andreozzi R, Caprio V. Water Res. 2009;**43**:2710-2716.
[17] Li L, Salvador PA, Rohrer GS. Nanoscale. 2014;**6**:24-42.
[18] Fujishima A, Honda K. Nature. 1972;**238**:37-38.
[19] Giordano N, Antonucci V, Cavallaro S, Lembo R, Bart JCJ. Int J Hydrogen Energy. 1982;**7**:867-872.
[20] Ghosh AK, Maruska HP. J Electrochem Soc. 1977;**124**:1516-1122.
[21] Sun B, Shi T, Peng Z, Sheng W, Jiang T, Liao G. Nanoscale Res Lett. 2013;**8**:1-8.
[22] Int J Photoenergy. 2012;**2012**:1-12.
[23] Yan-Qing C, Zhe L, Wang Qi, Yi Z, Qian X, Fang-Xia F. Acta Phys- ChimSin. 2012;**28**:1489-1496.

Acknowledgement
The authors would like to thank Centre of Innovative Nanostructures & Nanodevices of Universiti Teknologi PETRONAS for technical and financial support to make this research work feasible.

IOP Conf. Series: Materials Science and Engineering **348** (2018) 012008 doi:10.1088/1757-899X/348/1/012008

Morphological control of $La_{0.7}Sr_{0.3}Co_{0.2}Fe_{0.8}O_{3-\delta}$ and $La_{0.7}Sr_{0.3}MnO_{3-\delta}$ catalytic membrane using PEG-H_2O additive

R M Iqbal[1], S D Nurherdiana[1], D Hartanto[1], M H D Othman[2] and H Fansuri[1]

[1]Department of Chemistry, Institut Teknologi Sepuluh Nopember, Kampus ITS Sukolilo, Surabaya 60111, Indonesia
[2]Advanced Membrane Technology Research Centre (AMTEC), Universiti Teknologi Malaysia, 81310 Skudai, Johor, Malaysia
E-mail: h.fansuri@chem.its.ac.id

Abstract. Methane is the primary combustible component in non condensable part of natural gas. It is a promising source for syngas (CO and H_2) production by partial oxidation method. The conversion of methane to syngas by partial oxidation method needs a controlled amount of oxygen. Membrane which has asymmetric structure and selectively permeates oxygen can be used to supply just enough oxygen to the reaction. One pathway to the fabricate asymmetric membrane is phase inversion method with an addition of PEG to increase pore size. $La_{0.7}Sr_{0.3}Co_{0.2}Fe_{0.8}O_{3-\delta}$ (LSCF 7328) and $La_{0.7}Sr_{0.3}MnO_{3-\delta}$ (LSM 73) powder were synthesized by solid-state method and they were characterized by XRD. The green membrane was prepared by phase inversion method. A dope solution was made by mixing LSCF 7328 or LSM 73 powder with PEG and stirred them in NMP for 24 h. PESf was then added into the dope solution and the stirring was continued to another 24 h. The resulted dope solution was degassed by immersing the solution inside and conical flask in an ultrasonic bath to remove air bubbles. The degassed mixture was then casted by spreading it on a glass surface (with a thickness of 2 mm) followed by immersion in a water bath for 24 h to coagulate the degassed mixture. Membrane morphology was characterized by Scanning Electron Microscopy (SEM) while the decomposition temperature of the polymer binder was analyzed by Thermogravimetric Analyzer (TGA). The XRD results show that phase of LSCF 7328 and LSM 73 are similar to $LaCoO_3$ and $LaMnO_3$, respectively. It indicated that the perovskite synthesis was successful. SEM micrograph of membrane cross sections show that the green membrane have finger like pores and a dense layer. Pores also appear on top and bottom surface of the membrane. Based on TGA results, the highest weight lost of green membrane at 550-600°C which represents the decomposition of PESf binder.

1. Introduction

Indonesia has numerous natural resources such as natural gas which are promising to solve some problems in the energy sector. The primary hydrocarbon component in non condensable part of natural gas is methane and it can be converted to syngas as raw material to produce hydrocarbons or liquid fuels. Methane conversion to liquid fuels is known as Gas To Liquid (GTL) process which has two steps[1]. The first is methane conversion to syngas by, for example, partial oxidation [2]. The second step is Fischer tropsch reaction of the syn gas to hydrocarbons.

The conversion of methane to syngas by partial oxidation method needs a controlled amount of oxygen. If there are excessive oxygen, methane will be oxidized completely into CO_2 and H_2O, instead of partially oxidized to syn gas. In this case a membrane which selectively permeates oxygen can be used to supply just enough oxygen to the reaction. Teraoka et al reported that perovskite oxides are promising materials for such membrane to transfer oxygen where the oxygen flux can be controlled by modification of A and B site in ABO_3 perovskite lattice [3]. Perovskite oxides have general formula of ABO_3, where A site is occupied by a large cation such as alkaline, earth alkaline, and rare earth cation. Meanwhile B site is occupied by a smaller size cation like especially cation of transition metals [4].

Among many perovskite oxides, family of Sr^{2+} substituted $LaCoO_3$ (LSCF) and $LaMnO_3$ (LSM) attracted attention of many researchers due to their good oxygen flux. For an example, Tan et al reported the use of $La_{0.6}Sr_{0.4}Co_{0.2}Fe_{0.8}O_{3-\delta}$ as membrane material for oxygen separation and the result showed that the membrane has good flux and high selectivity to oxygen permeation [5]. The oxygen flux is higher when more La^{3+} was replaced by Sr^{2+}. However, at higher Sr^{2+} content, the perovskite react with CO_2 to form $Sr(CO)_3$ an reduce the oxygen flux. Therefore it is very important to limit the number of Sr^{2+} in such a way so the maximum Sr^{2+} substitution was limited to 30% $La_{0.7}Sr_{0.3}Co_{0.2}Fe_{0.8}O_{3-\delta}$ (LSCF 7328) and $La_{0.7}Sr_{0.3}MnO_{3-\delta}$ (LSM 73). In addition to the oxygen permeation, the membrane also have good electron conductivity [6]. The combination of good oxygen and electron conductivity makes the perovskite oxide can conduct oxygen without any electrical current.

As an oxygen supplyer to partial oxidation reaction, the oxygen flux of Sr^{2+}-LSCF and -LSM is a function of the membrane's thickness. Thinner membranes will have higher oxygen flux. the In order to compensate the reduction of oxygen flux in low Sr^{2+}-LSCF and -LSM, the membrane have to be made as thin as possible. An asymmetric membrane morphology, where a thin layer of membrane is supported on a thicker and porous support, is an alternative way to improve the oxygen flux without having to modify its chemical composition.

One promising way to fabricate asymmetric perovskite based membrane by phase inversion method, and it is combined with sintering technique [7]. Phase inversion method needs polymer as binder and pore structure can modify by addition of an additive to form the pore or increase pore size. Bakeri et al using water, methanol, ethanol, glycerol and acetic acid as the additive, the result showed membrane have finger like structure except for water additive produce membrane that has sponge like structure [8]. Mohamed et al also used water as an additive, increasing amount of water into dope solution would decrease finger like and increase the thickness of dense layer [9]. The disadvantages if used liquid as the additive were possible to vaporize when mixture process. If additive vaporize before phase inversion process, the membrane has low pore size or it would obtain the dense membrane. Another pathway to solve this problem were used the polymer as additive. Polyvinyl pyrrolidone (PVP) were reported by many researchers as an excellent additive to form the pore on perovskite membrane [5]. Besides PVP, another promising polymer additive was PEG, Saljoughi et al used PEG additive for modifying morphology and membrane structure of cellulose acetate [10]. The result showed, increasing of PEG loading or molecular weight into dope solution would increase pore size. Based previous research, in this work use combination PEG and water as the additive to modify morphology and membrane structure of LSCF 7328 and LSM 73.

2. Methodology

2.1 Materials

The raw material for LSCF 7328 and LSM 73 synthesis were p.a. grade of La_2O_3 (99.5%), $SrCO_3$ (99.9%), Co_3O_4 (99.5%), Fe_2O_3 (97%) and $MnCO_3$ from Merck. The flat membrane was made by using PESf (polyethersulfone) as the binder, NMP (N-Methyl-2-pyrrolidone) as the solvent, PEG 8000 as a pore modifier and water as nonsolvent (coagulant).

2.2 Synthesis of LSCF 7328 and LSM 73

LSCF 7328 7328 and LSM 73 was synthesized by a combination of mechanochemical and solid state method according to the method that was reported by Nurherdiana et al [11]. The product was characterized by XRD Panalytical using Cu Kα (λ = 1.5406 Å) which was generated by applying 30 mA current and 40 kV of voltage. The scan rate was $1^{o}.min^{-1}$ and step size was 0.02^{o}.

2.3 Preparation and Morphological Characterization of LSCF 7328 and LSM 73 Green Membrane

The green membrane was prepared by phase inversion method. A dope solution was made by mixing LSCF 7328 or LSM 73 powder with PEG and stirred them in NMP for 24 h. PESf was then added into the dope solution and the stirring was continued to another 24 h. The resulted dope solution was degassed by immersing it in an ultrasonic bath to remove air bubbles.

The degassed mixture was then casted by spreading it on a glass surface (with a thickness of 2 mm) followed by immersion in a water bath for 24 h to coagulate the degassed mixture. After coagulation step, the resulted membrane (green membrane) was dried to remove water. The morphology of the green membrane was characterized by SEM to analyze it's morphology and TGA-DTA to examine the decomposition of PESf and other non-organic content of the membrane. Table 1 shows the chemical composition of dope solution.

Table 1. Composition of Dope Solution

LSCF 7328 (wt%)	LSM 73 (wt%)	PESf (wt%)	NMP (wt%)	Additive (wt%)		Code
				PEG	H_2O	
52.1	-	6.7	41.2	-	-	A1
47.1	-	6.7	41.2	3.5	1.5	B1
47.1	-	6.7	41.2	4	1	C1
47.1	-	6.7	41.2	4.5	0.5	D1
47.1	-	6.7	41.2	5	-	E1
-	52.1	6.7	41.2	-	-	A2
-	47.1	6.7	41.2	3.5	1.5	B2
-	47.1	6.7	41.2	4	1	C2
-	47.1	6.7	41.2	4.5	0.5	D2
-	47.1	6.7	41.2	5	-	E2

3. Result and discussion

LSCF 7328 and LSM 73 were synthesized using a combination of mechanochemical and solid state method. The X-ray diffractogram (XRD) of the resulted perovskites are shown in Figure 1 and 2. The figure shows that diffractogram of LSCF 7328 is similar to $LaCoO_3$ (JCPDS 00-025-1060) while diffractogram of LSM 73 is similar to $LaMnO_3$ (JCPDS 00-035-1353). The similarities shows that the synthesis procedure successfully produce LSCF 7328 and LSM 73. In addition to the crystal structure, the diffractograms also shows that there is no impurities in the resulted perovskite.

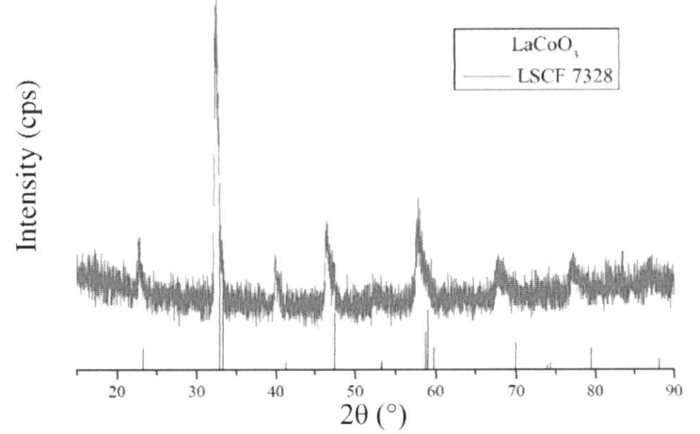

Figure 1. XRD Pattern of LSCF 7328

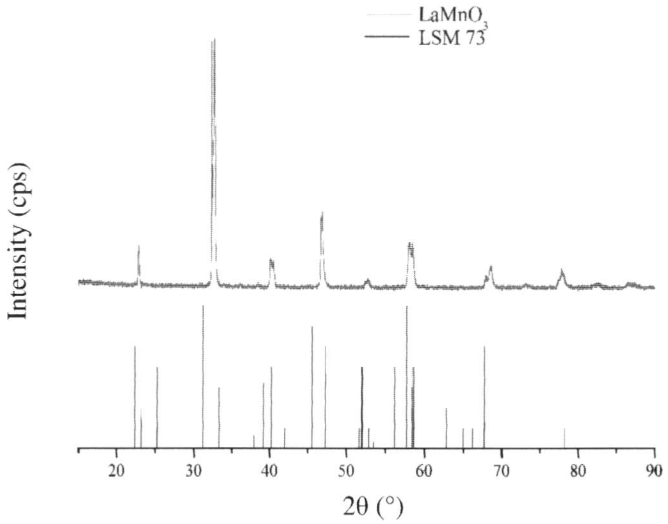

Figure 2. XRD Pattern of LSM 73

Further examination of diffractogram in Figure 1 revealed that specific diffraction peaks of LSCF 7328 was shifted to lower of 2θ. The partial substitution of Co^{3+} with a slightly larger Fe^{3+} and partial substitution of La^{3+} (1.5 Å) with larger Sr^{2+} (1.58 Å) cation makes the unit cell size larger than $LaCoO_3$. The increase of unit cell size also increase the interplanar distances of the perovskite crystal which was indicated by the shifts of all peaks to lower 2θ [1].
Phase inversion method was used to prepare LSCF 7328 and LSM 73 membranes. The advantage of the method is pore structure and size can be modified by modifying the diffusion rate of solvent (J1) and nonsolvent (J2). When J1=J2 the resulted membrane pores will be sponge like while when J2>J1 the pores will be finger like [12]. Figure 3 shows the mechanism of pore formation as a function of diffusion rate of solvent and nonsolvent.

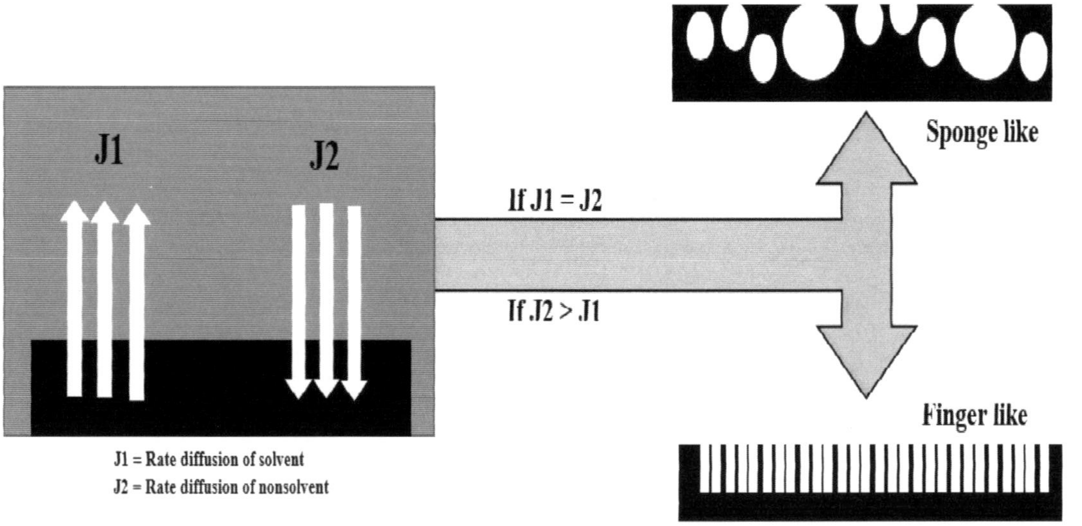

Figure 3. Pore formation mechanism

Table 2 and 3 show SEM Micrographs of LSCF 7328 and LSM 73 membrane with different loading of additive. In Table 2, it is shown that increasing PEG concentration in the dope solution gradually increase pore size of the membrane and each membrane has finger like pores. Membranes with finger like pores have better mechanical strenght than sponge like pores. It is also shown that each LSCF 7328 membrane has a thin layer of dense membrane on the top of membrane cross section.

In addition to the cross section, the top and bottom surface of the membrane was also identified by SEM micrograph. The top and bottom part of LSCF 7328 membrane have some small size pores which was probably formed during the diffusion of PEG into nonsolvent. The small pores, especially on the dense surface, may be reduced during the sintering process which will increase the densification of LSCF 7328 membrane [7].

Table 2. SEM Micrograph of LSCF 7328 Green Membranes

Sample Code	Bottom Surface	Top Surface	Cross Section
A1	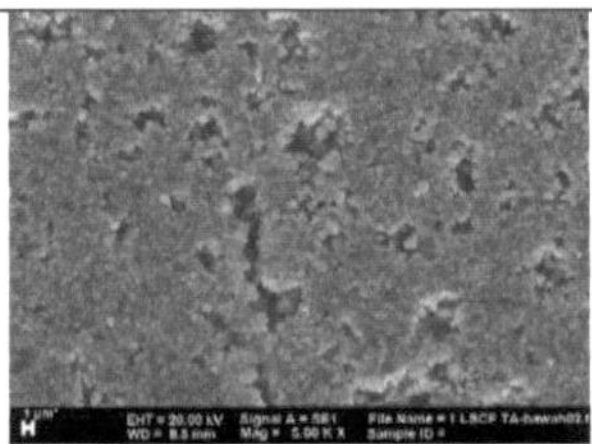	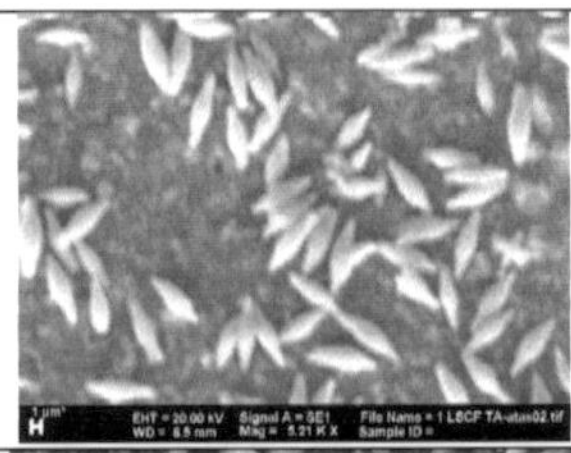	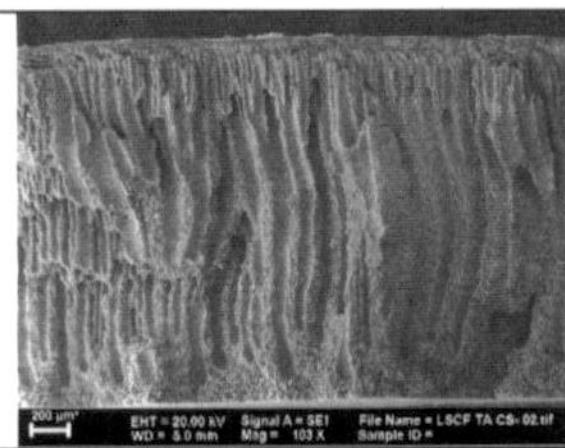
B1		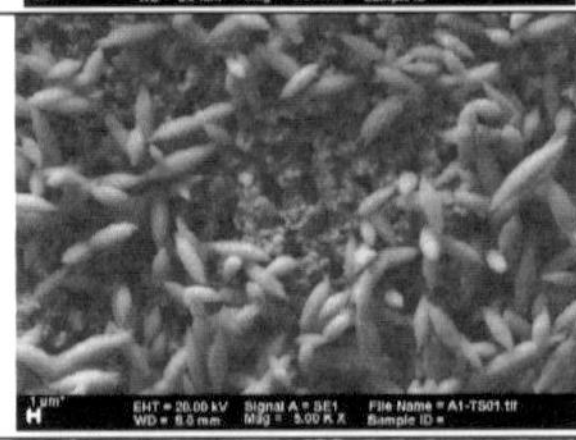	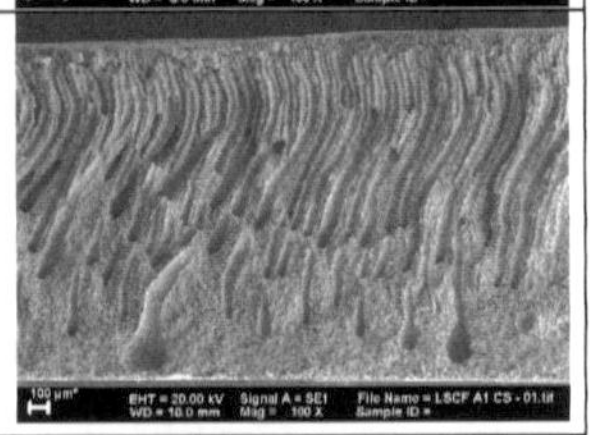
C1			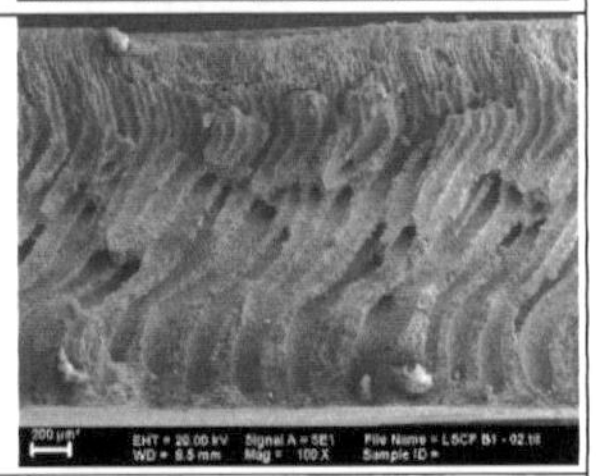
D1			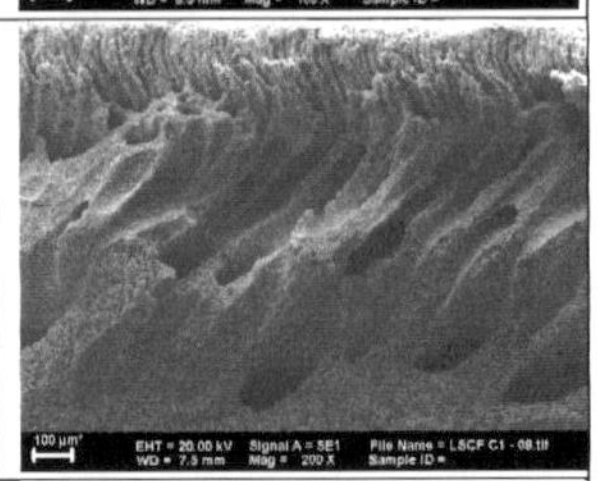
E1		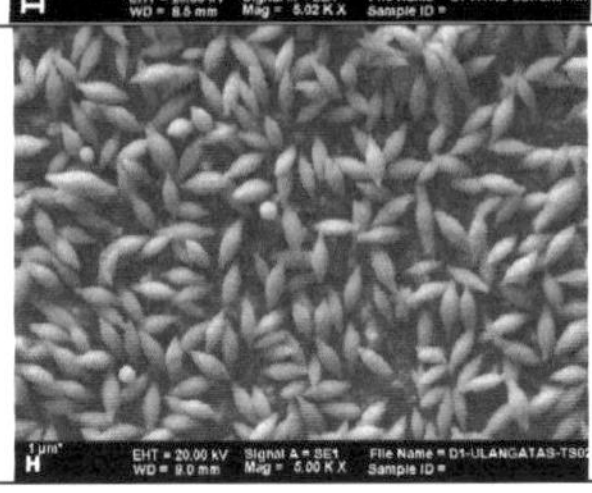	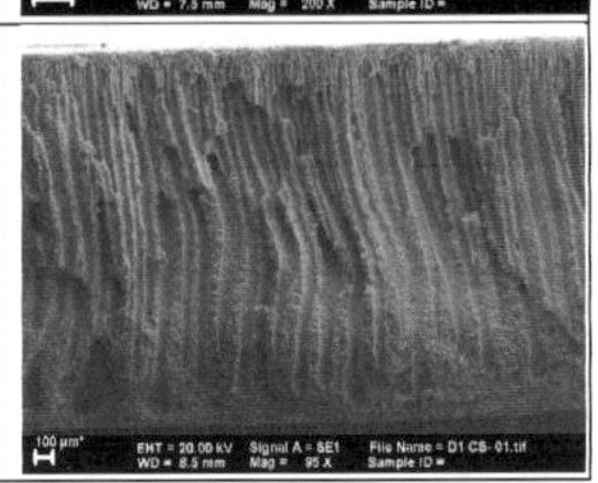

When comparing the pores on LSCF 7328 and LSM 73 membranes, the LSM 73 has less regular pores than LSCF 7328. It indicates that each powder has its own characteristic and properties which affect the pore formation during phase inversion process. Top and bottom surface of LSM 73 membrane showed the similar result to LSCF 7328 membrane where some small pores are observed. LSM 73 membrane also shows thicker dense layer than LSCF 7328. The thickness of dense layer will have effect on its oxygen flux and ultimately to the catalytic activity on partial oxidation of methane (POM) reaction [1]. Membranes with thicker layer of then dense part will have lower oxygen flux.

Table 3. SEM Micrograph of LSM 73 Green Membranes

Sample Code	Bottom Surface	Top Surface	Cross Section
A2			
B2			
C2	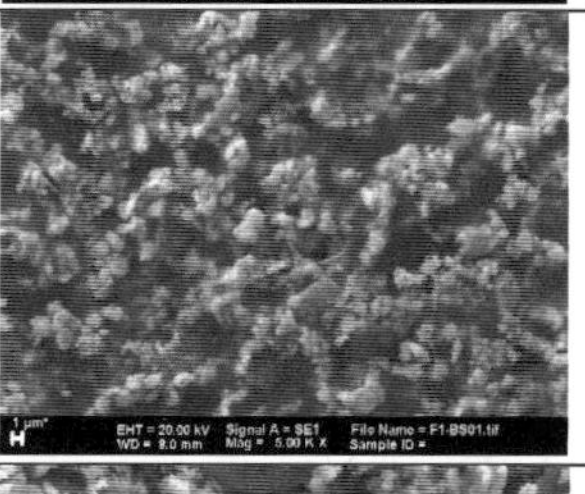	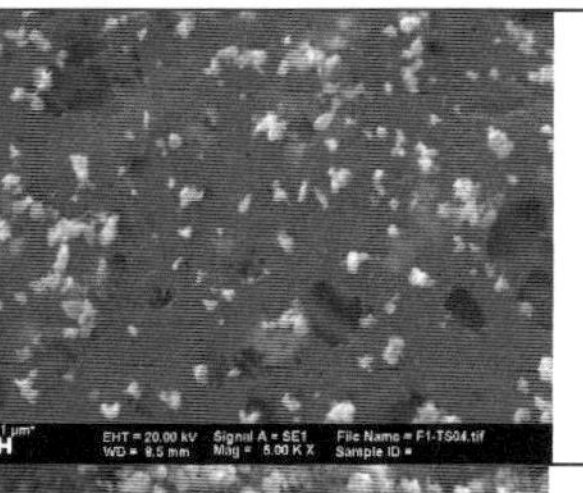	
D2		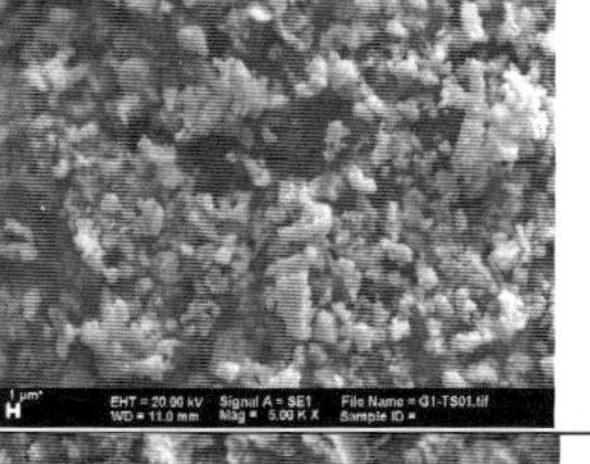	
E2			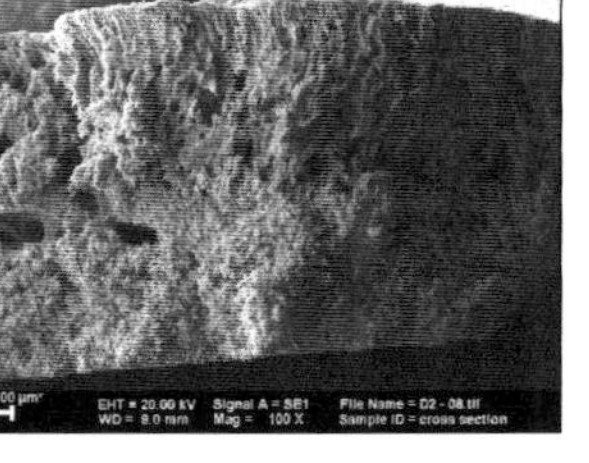

TGA analysis results of LSCF 7328 (LSCF-5) and LSM 73 (LSM-5) green membranes are shown in Figure 4. It is apparent that the highest lost weight of both green membranes occurs at around 550-600°C. Tan et al reported that the wight lost at 550-600°C is due to the decompotion of PESf. The weight lost at higher temperature might be due to thermal reduction of perovskites where some of their structural oxygen escape from the perovskites, leaving some oxygen vacancies. The reduction might be happening because the TGA analysis was carried in nitrogen atmoshpere that makes the oxygen partial pressure in the surrounding atmosphere was very low, similar to the reduction of titanate provskites that was reported by Rodenbücher et al. [13].

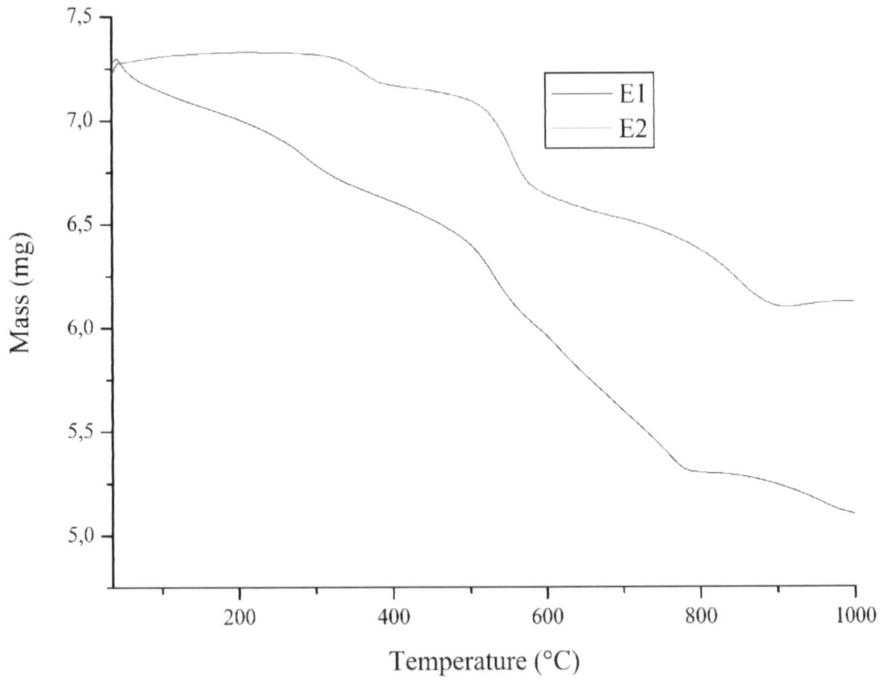

Figure 4. TGA Curve of LSCF 7328 (E1) and LSM 73 (E2) Green Membranes

4. Conclusion

LSCF 7328 and LSM 73 were successfully synthesized by solid state method. LSCF 7328 and LSM 73 has a similar structure with $LaCoO_3$ and $LaMnO_3$, respectively. Partial substitution to the lower size of ion caused shifted peak to the lower 2θ. Then, fabrication of each membrane use PEG-H_2O additive by phase inversion method obtained membrane that has finger like structure of their pores and thin dense layer. The higher PEG loading lead to increasing pore size of finger like and some pore with small size appear on the bottom and top surface of the membrane. Thermogravimetric analyzer result exhibit decomposition of PESf binder around 550-600°C and increasing temperature would increase densification of LSCF 7328 and LSM 73 particle.

5. References

[1] Fansuri H, Syafi MI, Romdoni S, Masyitoh AD, Utomo WP, Prasetyoko D, Widyastuti N and Murwani IK 2017 *Advanced Materials Letters* 8 799-806
[2] Meng B, Zhang H, Jingcang Q, Tan X, Ran R and Liu S 2015 *Separation and Purification Technology* **147** 406-413
[3] Sunarso J, Baumann S, Meulenberg WA, Liu S, Lin YS and Diniz JC 2008 *Journal of Membrane Science* **320** 13-41
[4] Côrte RV, Conceição L and Souza MM 2013 *Ceramic International* **39** 7975-7982
[5] Tan X, Liu N, Meng B and Liu S 2011 *Journal of Membrane Science* **378** 308-318
[6] Takahashi Y, Kawahara A, Suzuki T, Hirano M and Shin W *Solid State Ionic* **181** 300-305
[7] Setyaningsih EP, Machfudzoh M, Utomo WP and Fansuri H *Indonesia Journal Chemistry* **16** 20-24
[8] Bakeri G, Matsuura T and Ismail AF *Journal of Membrane Science* **383** 159-169
[9] Mohamed MH, Othman MHD, Mutalib MA, Rahman M, Jaafar J and Ismail AF 2016 *International Journal Applied Ceramic Technology* **809** 799-809
[10] Saljoughi E, Amirilargani M and Mohammadi T *Desalination* **262** 72-78
[11] Nurherdiana SD, Scholichah N, Iqbal RM, Sahasrikirana MS, Utomo WP, Akhlus S, Nurlina N and Fansuri H 2017 *Key Engineering Materials* **744** 399-403

[12] Mulder M 1992 *Basic Principle of Membrane Technology* Kluwer UK 86
[13] Rodenbücher, C, Meuffels, P., Speier, W., Ermrich, M, Wrana, D, Krok, F, and Szot, K 2017 *Phys. Status Solidi RRL* 2017, 00, 1700222

Acknowledgments
This work was financially supported by Institut Teknologi Sepuluh Nopember under Lab Based Education Grant 2017 with Contract No : 820/PKS/ITS/2017. We thank to Advanced Membrane Technology (AMTEC) – Universiti Teknologi Malaysia for providing research facilities.

IOP Conf. Series: Materials Science and Engineering **348** (2018) 012009 doi:10.1088/1757-899X/348/1/012009

Characterization of water based nanofluid for quench medium

Kresnodrianto [1, a)] , S. Harjanto[1, b)] , W.N. Putra [1, c)], G. Ramahdita [1, d)], S.S. Yahya[1], E.P. Mahiswara [1]

1 Material and Metallurgy Departement, Faculty of Engineering, Universitas Indonesia, Depok, Indonesia
a) kresnodrianto@ui.ac.id
b) sri.harjanto@gmail.com
c) wahyuaji@metal.ui.ac.id
d) ghiska@metal.ui.ac.id

Abstract- Quenching has been a valuable method in steel hardening method especially in industrial scale. The hardenability of the metal alloys, the thickness of the component, and the geometry is some factors that can affect the choice of quench medium. Improper quench media can cause the material to become too brittle, suffers some geometric distortion, and undesirable residual stress that will cause some effect on the mechanical property and fracture mechanism of a component. Recently, nanofluid as a quench medium has been used for better quenching performance and has been studied using several different fluids and nanoparticles. Some of frequently used solvents include polymers, vegetable oils, and mineral oil, and nanoparticles frequently used include CuO, ZnO, and Alumina. In this research, laboratory-grade carbon powder were used as nanoparticle. Water was used as the fluid base in this research as the main observation focus. Carbon particles were obtain using a top-down method, whereas planetary ball mill was used to ground laboratory grade carbon powder to decrease the particle size. Milling speed and duration were set at 500 rpm and 15 hours. Field Emission Scanning Electron Microscope (FE-SEM), and Energy Dispersive X-Ray (EDX) measurement were carried out to determine the particle size, material identification, particle morphology, and surface change of samples. Nanofluid was created by mixing percentage of carbon nanoparticles with water using ultrasonic vibration for 280s. The carbon nanoparticle content in nanofluid quench mediums for this research were varied at 0.1%, 0.2%, 0.3%, 0.4, and 0.5 % volume. Furthermore, these mediums were used to quench JIS S45C or AISI 1045 carbon steel samples which austenized at 1000°C. Hardness testing and metallography observation were then conducted to further check the effect of different quench medium in steel samples. Preliminary characterizations showed that carbon particles dimension after milling was still in sub-micron stage, hundreds of nanometres to be precise. Therefore, the milling process parameters are needed to be optimized further.

1. Introduction

Heat Treatment and quenching have been one of the important process of steel hardening in industrial scale. Quenching refers to the fast cooling of a steel. In quenching, an austenized steel is immersed in a liquid, called quench medium or quenchant[1]. Quench medium that commonly used are water, brine, polymers, oils, salts, and gases which have their own cooling characteristics and results. Improper quench media can cause the material to become too brittle, suffers some geometric distortion, and undesirable residual stress that will cause some effect on the mechanical property and fracture

mechanism of a component. The result of quenching, is highly effected by the cooling rates of quench media. In quenching, to get a good result, a cooling rate needs to be equal or higher than the critical cooling rate for martensite formation[2]. Usually, for components that has variation on their thicknes, we use quench medium with low cooling rates to avoid excessive residual stress, since a rapid cooling has a high possibility of the occurance of cracks, especially on the critical parts of the components.
Nanofluid refers to a fluid containing a small amount dispersed nano-sized (< 100 nm) solid nanoparticles. It has been proved that the nanoparticles will enhance thermal conductivity and improve thermal performance of heat transfer of the base fluid[3]. Studies on the performance of nanofluids with various types of particles has been done by many researchers[1, 4-11].
The present study aims to compare the quenching of carbon steel S45C using deionized (DI) water-based nanofluids with laboratory-grade carbon powder as nanoparticles. The samples were heated at 1000 ^{o}C for an hour and then quenched. The effect of nanofluid concentraion is investigated to illustrate an optimal ratio of nanoparticle that is used in achieving the desired material properties.

2. Materials and methods

S45C carbon steel bar was purchased from a local steel distributor. The chemical composition of the carbon steel sample was determined as given in Table 1.

Table 1. Chemical composition of S45C carbon steel (%)

Fe	C	Si	Mn	P	S	Cr	Mo
98.3	0.47	0.287	0.718	0.0261	0.005	0.028	0.005
Ni	**Al**	**Co**	**Cu**	**Nb**	**Ti**	**V**	**W**
0.005	0.02	0.003	0.018	<0.002	0.035	<0.002	<0.01
Pb	**Sn**	**B**	**Ca**	**Zr**	**As**	**Bi**	
0.025	<0.002	0.0013	0.0005	<0.002	0.008	<0.03	

This study started by preparation of specimen of medium carbon steel about 30 samples. The samples were cut in the dimension of 15mm x 10mm x 10mm. Among this samples, non-heat treated were prepared as baseline comparison. Steel samples were pre-heated at 600 ^{o}C and then austenized at 1000 ^{o}C and holding it for an hour. Those samples were then quenched in water-based nanofluid. A vickers hardness testing is conducted to measure the hardness of the samples. In the hardness testing, a force of 300 gf is applied to press the diamond indentor into the samples, and the force is maintained for 10 s. Finally, the hardness of the materials can be calculated by the size of the indentation. To obtain the average hardness, five different locations on each sample are selected to measure the hardness.
For nanofluids preparation, laboratory-grade carbon powder was used to synthesized oil-based nanofluids by two-step method. The nanofluids of 0.1%, 0.2%, 0.3%, 0.4% and 0.5% volume fraction of particles were prepared. The nanoparticles were first produced by high energy milling. The second step involves dispersing this nanopowder into the base fluid with the help of intensive ultrasonic agitation. The two-step method was more economical than the one-step method to produce nanofluids commercially and also considered the best method in dispersing carbon particles in a base fluid[12]. This method is suitable forwide range of particles such as oxide particles and carbon nanotubes and it is attractive to industry because it is simple for nanofluid preparation[13].
Nanoparticles synthesis was carried out at 500 rpm for 15 hours in the high energy ball mill. The nanoparticles were weighted according to the weight of each quenching medium based on % volume. After this step, water was slowly added to 100 ml beaker glass which filled of nanoparticles powder. In order to increase the stability of solutions, each nanofluid suspension was sonicated in an ultrasonic apparatus for 280 seconds before the quenching experiments.
Microstructure examination were carried out for both treated and untreated samples. Each sample was prepared using standard metallography procedure. The phase of the specimens were made visible by etching using nital containing 2% Nitric acids and 98% methylated spirit on the polished surfaces. Then the specimen was immersed into etching media for 5 seconds. Microscopic examination of the etched surface of various specimens was taken using a metallurgical microscope.

3. Results and discussion

FE-SEM imaging was conducted to examine the everage particle size of the laboratory-grade carbon powder before milling (Figure 1). From this examination, the average size of the powder is 15,03767 μm.

Figure 1. FE-SEM image of laboratory-grade carbon powder with (a) 500x magnification and (b) 1000x magnification.

Figure 2. EDS layered image of laboratory-grade carbon powder.

Table 2. Result of EDS for chemical composition (wt%) of laboratory-grade carbon powder

Laboratory-grade carbon powder	C	Cu
	99.9	0.1

Figure 2 showed the EDS layered micrograph and elemental distribution of laboratory-grade carbon powder. The elemental distribution confirms that the powdes is mainly consisy of carbon. The exact composition of the powder can be seen in Table 2. The laboratory-grade carbon powder consist of 99.9 wt% carbon and some impurities of 0.1 wt% Cu.

The microstructure of the non-heat treated sample is shown in Figure 3. Figure 3 showed that the non-heat treated sample mainly consist of ferrite and pearlite. The ferrite structure is the white area, while the darker area is pearlite.

IOP Conf. Series: Materials Science and Engineering **348** (2018) 012009 doi:10.1088/1757-899X/348/1/012009

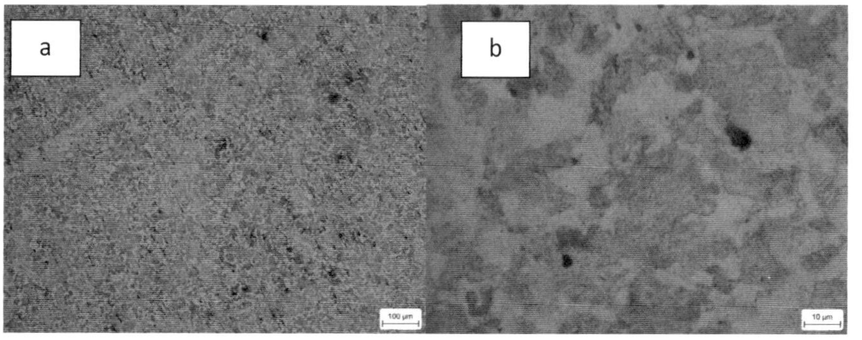

Figure 3. Microstructure of non-heat treated sample (a) 100x magnification (b) 1000x magnification.

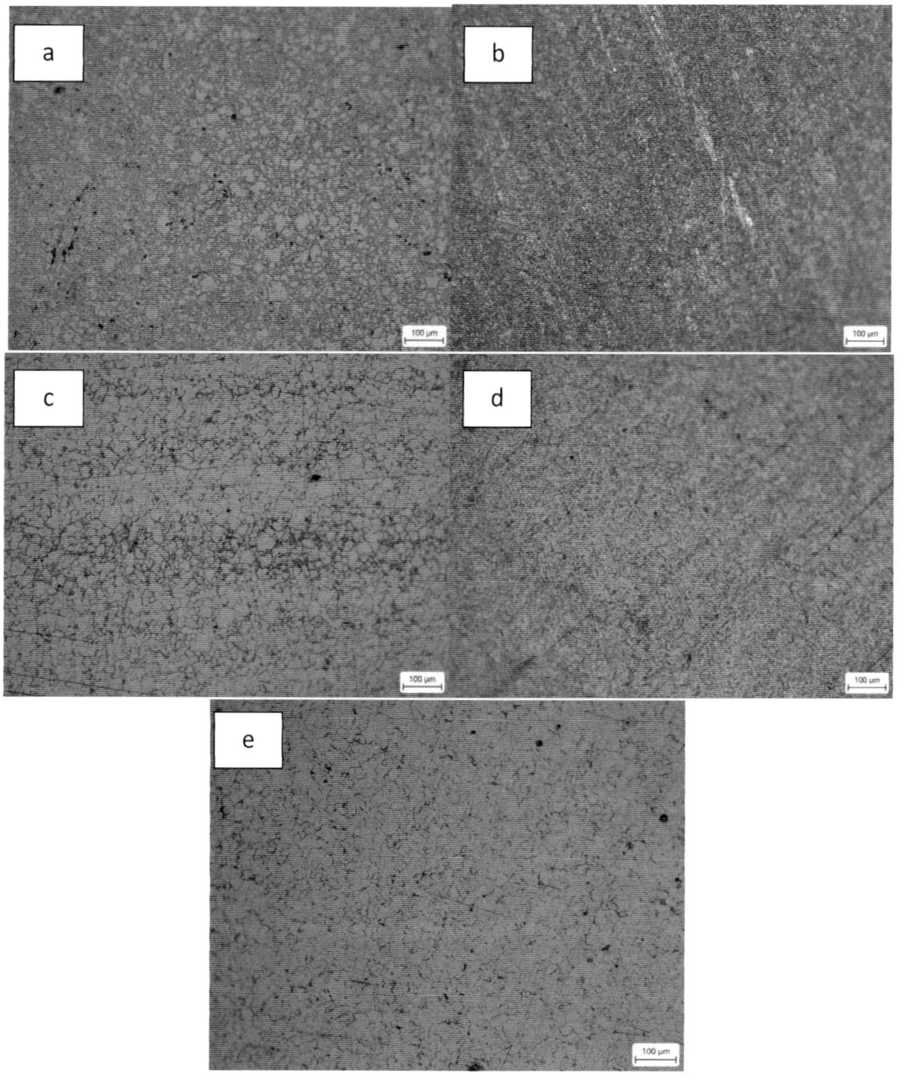

Figure 4. Microstructure of S45C after quenching in water-based laboratory grade nanofluid (a) 0.1% vol. 100x magnification ,(b) 0.2% vol. 100x magnification, (c) 0.3% vol. 100x magnification, (d) 0.4% vol. 100x magnification, and (e) 0.5% vol. 100x magnification.

Figure 4 shows the microstructure of S45C after quenching in water based laboratory grade nanofluid. Martensite structure is observed in the samples which were quenched in 0.2% vol. Nanodluid, which can be observed as the one that has the densest martensite structure.The rest also exhibit martensite structure, but in a different amount. This anomaly happens because the agitation by ultrasonic is poorly done, so the nanoparticle was not dispersed fully which effect the nanofluid performance.
The result of hardness Vickers testing of the sample is shown in Figure 5. Figure 5 show that the first samples that is quenched with DI Water exhibit the hardness number of 675 HV. Samples that quench in water-based nanofluid gives a unique behavior, it's hardness rises until 0,2% and then it drops. The hardness The maximum hardness in water-based nanofluid is obtained by 0,2% vol. nanofluid with 885,34 HV. The hardness value is linear to it's microstruture which shows the most martensite structure. This result shows that carbon nanoparticles inside the fluid can enhance the fluid severity with appropriate concentration. This happen due to transformation of martensite with rapid cooling. Transformation of martensite is driven by nanoparticles in the fluid which is having higher thermal conductivity than in normal fluid. The heat transfer enhancement phenomenon can be illustrated as the formation of nanoparticle embedded surface due to scattered deposition of nanoparticles. As a result, it increases the effective surface area of heat transfer[14].

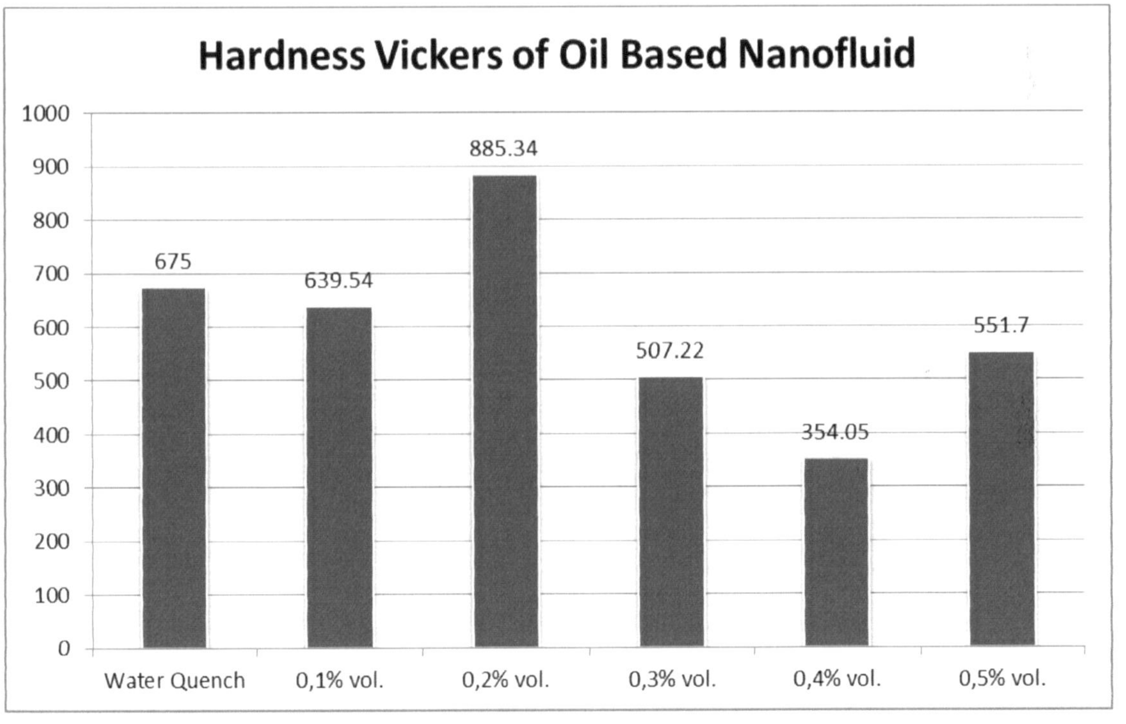

Figure 5. Result of hardness vickers testing.

4. Conclusion

The effect of laboratory-grade carbon in water as quench medium was experimentally investigated using medium carbon steel, JIS S45C as specimen. For this purpose, various concentration of nanofluids were prepared with two step method. The variation of concentration is 0.1%, 0.2%, 0.3%, 0.4% and 0.5% volume. The microstucture and hardness value shows that volume fraction of nanoparticles used in nanofluids considerably affected the quenching process. It is observed that after quenching with water-based nanofluid, the microstructure mainly consist of martensite. The maximum hardness is observed in 0,2% vol. which is 885,34 HV. The hardness number is decreasing in 0,3% vol. and increasing again in 0,4% vol and decreasing again in 0,5% vol. This anomaly happened because of the poor mixing process using ultrasonic vibration, that causes agglomeration between the carbon nanoparticles. The hardness value is higher than cooling without nanoparticles added. This concludes that nanoparticles inside the fluid can enhance the fluid severity with appropriate concentration.

5. Reference

[1] Nayak, Vignesh U. Prabhu, Narayan. K. 2016. "Wetting Behavior and Heat Transfer of Aqueous Graphene Nanofluids.
[2] J. Grum, S. Bozic, and M. Zupancic, Influence of Quenching Process Parameters on Residual Stresses in Steel, J. Mater. Process. Technol., 2001, 114(1), p 57–70
[3] Dhinesh K D and Valan A A. 2016. *A review on preparation, characterization, properties and applications of nanofluids*. (India: Elsevier) 21-40.
[4] Xie H, Wang J, Xi T, Liu Y, Ai F and Wu Q, "Thermal conductivity enhancement of suspensions containing nanosized alumina particles", (J Appl Phys, 2002) vol. 91:4568-4572.
[5] S. Lee, S.U.S. Choi, S. Li, and J.A. Eastman, "Measuring thermal conductivity of fluids containing oxide nanoparticles", (Journal of Heat Transfer, 1999) vol. 121, pp. 280-289.
[6] X. Wang, X. Xu, and S.U.S. Choi, "Thermal conductivity of nanoparticle–fluid mixture," (Journal of Thermophysics and Heat Transfer, 1999) vol. 13, pp. 474-480.
[7] J.A. Eastman, S.U.S. Choi, S. Li,W. Yu, and L.J. Thompson, "Anomalously increased effective thermal conductivities of ethylene glycol-based nanofluids containing copper nanoparticles," (Applied Physics Letters, 2001) vol. 78, no. 6, pp. 718-720.
[8] Ramesh, Gopalan and Narayan Kotekar Prabhu, "Review of thermo-physical properties, wetting and heat transfer characteristics of nanofluids and their applicability in industrial quench heat treatment", (Nanoscale Research Letters, 2011) (15 pages)
[9] Xuan Y and Li Q, "Heat transfer enhancement of nanofluids", (Int J Heat Fluid Flow, 2000) vol. 21 pp 58-64.
[10] Kim, H., DeWitt, G., McKrell, T., Buongiorno, J., and Hu, L.W., "On the quenching of steel and zircaloy spheres in water-based nanofluids with alumina, silica and diamond nanoparticles", (International Journal of Multiphase Flow, 2009) vol. 35, pp 427–438.
[11] Khoshmehr,Habibi, Ahmad Saboonchi,and Mohammad Behshad Shafii, "The quenching of silver rod in boiling carbon nano tube-water nanofluid", (International Journal of Thermal Sciences, 2014) pp 95-104
[12] A.M. Rashidi, M.M. Akbarnejad, A.A. Khodadadi, Y. Mortazavi, A. Ahmadpour, Single-wall carbon nanotubes synthesized using organic additives to Co-Mo catalysts supported on nanoporous MgO, Journal of Nanotechnology 18 (31) (2007). No.315605
[13] G. Paul, P.K. Das, and I. Manna, "Rewetting of vertical pipes by bottom flooding using nanofluid as a coolant", (Int. J. Heat Mass Transf., 2015) vol. 137 (9 pages).
[14] N. Patra, V. Gupta , R.Singh, R.S. Singh, P. Ghosh and A Nayak, "An experimental analysis of quenching of continuously heated vertical rod with aqueous Al2O3 nanofluid", (Resource-Efficient Technologies,2017) (8 pages)

Acknowlegment

Authors would like to thank the Directorate of Research and Public Services, Universitas Indonesia for financial support and project administration respectively to conduct this research under PITTA Project on 2017 fiscal year.

Study of corrosion behavior on the addition of sodium citrate in nickel electroplating on SPCC steel using EIS

R Riastuti[1], C Ramadini[1], S T Siallagan[1], A Rifki[1], and F Herdino[1]

[1]Metallurgy and Materials Department, Faculty of Engineering Universitas Indonesia, Kampus Baru UI Depok, 16424 – Indonesia
E-mail: riastuti@metal.ui.ac.id

Abstract. The addition of sodium citrate to nickel electroplating process as additive is useful for refining the grain size of nickel deposit. The refining of grain size in nickel deposit as coating layer can improve surface performance, one of which corrosion resistance. This paper aims to investigate the effect of sodium citrate addition as grain refiner to promote corrosion resistance on SPCC steel. This experiment used Watt's Bath solution of $NiSO_4$ 300 g /L, $NiCl_4$ 45 g/L, H_3BO_3 60 g /L, wetting agent 0.2 cc/ L. Sodium citrate was added in composition of 45g/L and 60g/L. Nickel were deposited by direct current using current density on 6 A/dm^2 at the acidity level of 5 for 30 minutes by keeping the operating temperature stable at 50°C. The grain size of nickel deposit was observed through Optical Microscope and Atomic Force Microscope (AFM). The corrosion behavior of SPCC was observed by linear polarization and Electrochemical Impedance Spectroscopy (EIS) methods using 3% NaCl solution. Based on the research, the addition of sodium citrate as grain refiner will increasing corrosion resistance on SPCC steel from 0.35 to 0.05 mm/year.

1. Introduction

SPCC steel has poor corrosion resistance, it could be easily oxidized in atmospheric condition. In industry, this type of steel is widely used beacuse of its very good mechanical properties [1]. Preventif treatment is needed to prevent the failure because of its poor corrosion resistance. One of the most easiest method is using electroplating process to make a layer that can prevent the surface of steel interacted directly with the atmosphere [2]. Metal whose reduction potential lower than the substrate could be use in electroplating process [3]. Nickel is either metal whose has potential reduction lower than ferrous. Nikel electroplating itself is a process to deposited nickel onto suface of substrate material with chemical reaction under direct current electricity [4],[5].

Nanocrystalline material is an important material that used by industry these days because of its good chemical properties. This type of material could be use as subject material in electroplating method. In present days, there are many reasearches focus on improving the performance of this material, but there is a plenty of researches that have been focused on its corrosion behavior when this material use as coating [6]. To improve the corrosion resistance of nickel nanocrystalline, refining grain size is one of the method that can used. Its corrosion resistance will increase corresponded to finer grain size of electroplated nickel [7]. This increasing of the behavior is because the density of grain boundary increased so that the passive layer could be form shortly [8].

Refining of the grain has been demonstrated with saccharine as brightener [9]. The utillization of saccharine as brightener could make intergranular embrittlement [10]. Because of that, these days has been used another compound to replacing saccharine. One of them is sodium citrate, it has been demonstrated in patent WO 1994012695 [11]. Although sodium citrate has been demonstrated in a

IOP Conf. Series: Materials Science and Engineering **348** (2018) 012010 doi:10.1088/1757-899X/348/1/012010

patent and has lower price than saccharine, there has not been much research which study sodium citrate as grain refinement and its mechanism [12].
This present work is aimed to study the corrosion resistance on SPCC steel with electroplated nickel with addition of sodium citrate. Corrosion resistance of this layer will be measured using linear polarization method and Electrochemical Impedanca Spectroscopy (EIS). Atomic Force Microscope topography was used to measure the roughness.

2. Experimental

In this experiment was used Steel Plate Cold Rolled Coiled (SPCC) with a thickness about 1.6mm as cathode and carbon as anode during electroplating process. The chemical composition of SPCC was determined by Optical Emission Spectroscopy (OES) in Table 1.

Table 1. Chemical Composition of SPCC Steel.

Material	Chemical Composition (%)			
	C	Mn	P	S
SPCC	0.0448	0.181	0.0107	0.0076

Mechanical and chemical preparation were performed on SPCC strips (10mm x 35mm x 1.6mm). Surface preparation for the steel was done by manually polishing with amery papers in 60,120, and 180 grade. It was done to remove oxide layer in the surface which form because of the oxidation process with the atmosphere. After that, SPCC strips degreased using 10wt% NaOH at temperature of 70°C for 11 minutes and pickling using 30 wt % HCl at room temperature for 10 minutes. The last preparation step is etching using H_2SO_4 10wt% for 4 minutes..
Electroplated Ni were prepared by direct current (DC) electroplating from an electrolyte containing nickel sulphate (about 300 g/L), nickel chloride (about 30 g/L) and boric acid (about 30 g/L) at temperature of 50°C and pH 5 for 20 minutes. The current density was used at 6 A/dm^2 which is obtained based on hull cell test. This paper will investigated the corrosion behavior of SPCC steel in three condition that are a) plating Ni, b) plating Ni + 45g/L of sodium citrate and c) plating Ni + 60g/L of sodium citrate.
For the electrochemical testing, the sample was first mounted with epoxy where 1 cm^2 of its surface was exposed to electrolyte and the backside one was providing electrical connection. The electrochemical measurements were carried out in a three-electrode cell. SPCC coated with Ni was used as the working electrode, a platinum wire was used as the counter electrode, and Silver-Silver Chloride (SSC) was used as the reference electrode. The electrochemical measurement was analyze using AUTOLAB potentiostat/galvanostat system. Before all the measurements, the sample was immersed into electrolyte for 120 second to stabilize the open circuit potential (OCP). The aqueous solution containing 3wt.% NaCl was used as electrolyte in electrochemical cell and pH of electrolyte was measured at 6. The analysis of linear polarization result was analyzed using software NOVA 1.11.2.
The linear polarization method was used to determine the value of corrosion potential (E_{corr}) and corrosion current density (i_{corr}). And the potentiodynamic anodic polarization curves were also acquired based on this method. The linear polarization resistance was determined by performing linear sweep voltammetry (LSV) at a scanning rate of 10 mV/s. Scan direction was determined by start potential at -0.2 V and stop potential at 1.0 V. The measurement of EIS was conducted at the open circuit potential. The applied potential was 10 mV and the frequency ranged from 0.01 Hz to 100 kHz.
Physical characterization of the samples consisted of roughness of the grain result from electroplating process. Atomic force microscope topography was used to measure the roughness of each electroplated nickel layer.

3. Results and discussion

3.1. Effect ofsodium citrate on grain size

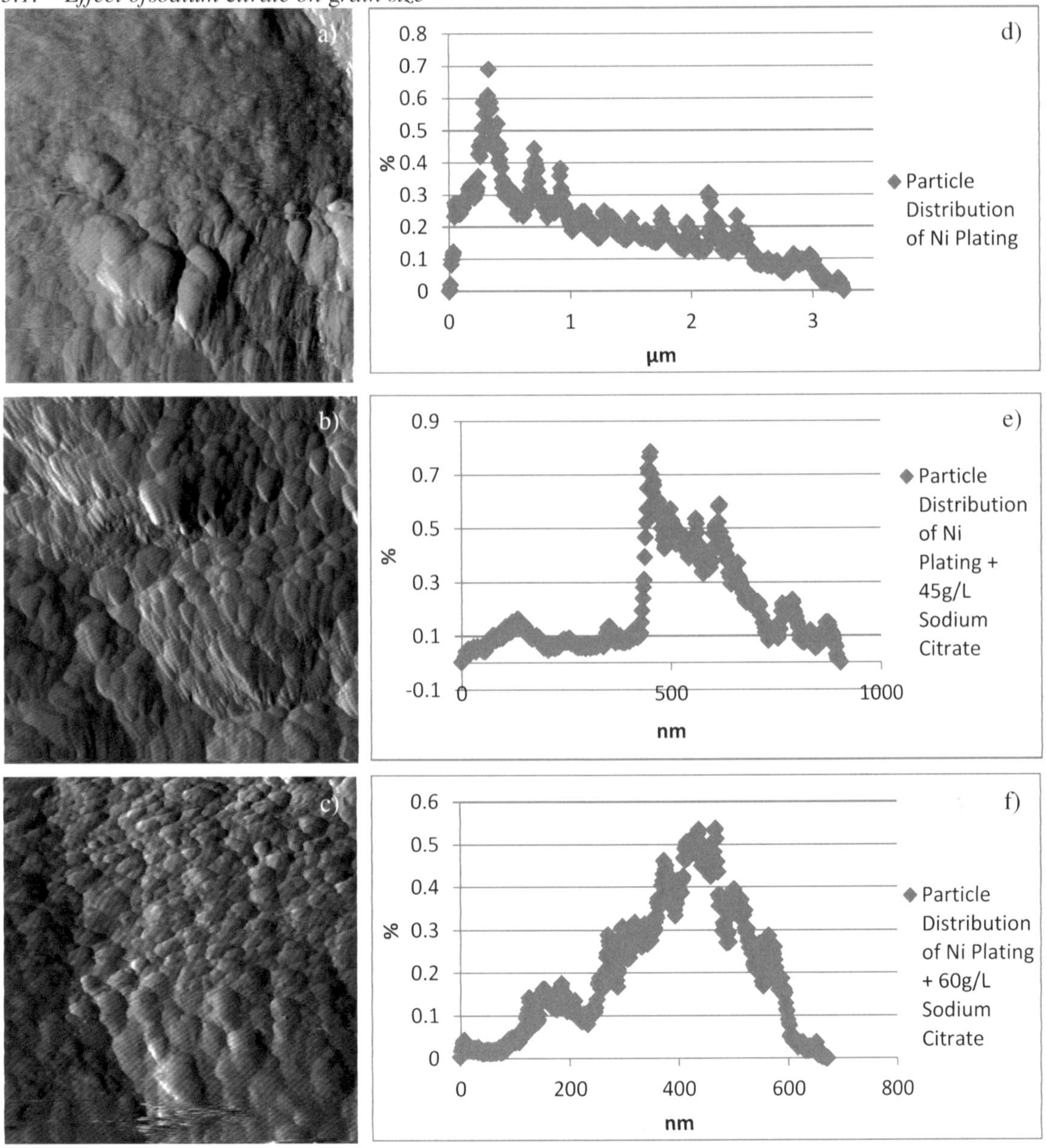

Figure 1. AFM topographic measurement for a) Ni plating, b) Ni plating + sodium citrate 45 g/L, c) Ni plating + 60 g/L; and particle distribution graphic for for d) Ni plating, e) Ni plating + sodium citrate 45 g/L, f) Ni plating + 60 g/L

Figure 1 a), b), and c) shows the result of AFM topographic measurement for all samples. Based on Figure 1, there is differences of the surface roughness which clearly show that all of samples have different range of grain size. Sample with the finest grain is sample electroplated nickel with the addition of 60 g/L sodium citrate.

Figure 1 d), e), and f) shows particle distribution graphic of all samples. Based on those graphics, grain size of electroplated nickel without additive is micrometer sized with the biggest size is about 3µm. Electroplated nickel with additive has finer grain size, it is nanometer sized with the biggest size of the addition of 45 g/L sodium citrate is about 900nm and the biggest size of the addition of 60 g/L sodium citrate is about 660nm. It can be concluded the addition of sodium citrate affected the grain size of electroplated nickel. The finest grain size is when the addition up to 60 g/L.

3.2. Effect of sodium citrate addition on corrosion resistance

3.2.1 Linear polarization measurement

Figure 2 shows polarization curve of the resulft of polarization measurement using linear polarization method. It shows that electroplated nickel with the addition of sodium citrate has more negative potential and the addition of 60 g/L sodium citrate has the most negative corrosion potential of all. More negative potential in the polarization measurement with 3wt.% NaCl shows that the material has better corrosion resistance [7].

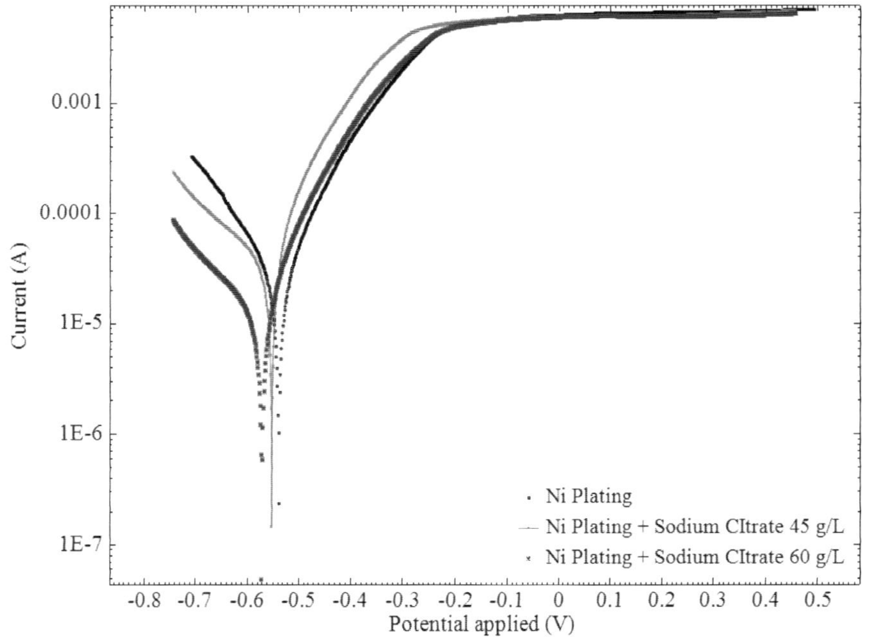

Figure 2. Polarization curve of electroplated nickel on SPCC steel with the addition and without the addition of sodium citrate as additive in 3wt% NaCl solution.

The result of polarization measurment can be seen in Table 2. Corrosion potensial (E_{corr}) of electroplated nickel on SPCC steel become more negative from -538.3 to -571.7 mV vs SSC along with the decreasing of current density from 30.4 to 4.34 μA cm^{-2}. The corrosion rate also decreased from 0.35 to 0.05 mm/year. The indication of decreasing potensial corrosion, current density, and corrosion resistance is the presence of sodium citrate as additive [7], [13], [16]. The addition of sodium citrate in electroplating process will refine grain structure of nickel layer on surface of the substrate. Refining of the grain size developed by transition of grain orientation that will inhibit the grain growth. This transition occur because of the changing of the surface energy [9]. Besides that, the addition sodium citrate to produce finer grain size affected to decreasing value of tafel slope (β_a and β_c).The lower value of tafel slope with decreasing grain size indicates has lower corrosion. This relates to the Stearn Geary equation:

$$i_{corr} = \frac{\beta a.\beta c}{2.303 \times Rp(\beta a + \beta c)} \quad (1)$$

where i_{corr} is the current density of corrosion and R_p is the polarization resistance. Based on the equation, the decrease of $\beta_a$and β_c will decrease the corrosion rate.
In the NaCl solution, the corrosion potential (E_{corr}) of the nickel layer will decrease as the corrosion resistance observed in Figure 2. The potential decrease is indicated because of the influence of the presence of the catalyst on the hydrogen reduction process in the form of a defect in the crystal lattice on the steel surface. It has been observed that in the presence of trapped hydrogen at dislocations, grain boundaries, and vacancies or pores are able to alter the kinetics of hydrogen evolution [19].

Table 2. Results of tafel extrapolation and linear polarization experiments of Ni plating in 3wt% NaCl solution.

Sample	β_a (mV/dec)	β_c (mV/dec)	E_{corr} (mV)	i_{corr} ($\mu A.cm^{-2}$)	Corrosion Rate (mm/year)
Ni Plating	141.7	132.5	-538.3	30.4	0.35
Ni Plating + Sodium Citrate 45 g/L	93.3	30.8	-553.4	12.7	0.15
Ni Plating + Sodium Citrate 60 g/L	59	51.2	-571.7	4.34	0.05

In the polarization curve of all the samples, there is passive condition when the line being steady state towards the current density. This passive condition occur beacuse of the apperance of passive layer on electroplated nickel on surface of the substrate. This passive layer could be form easier when the electroplated nickel has nano grain size particle beacuse it has bigger grain boundary density. This is also an indication of increasing corrosion resistance of nickel electroplated layer [8].

3.2.2 Electrochemical impedance spectroscopy (EIS) measurement
The Nyquist impedance plots for Ni plating with no additive and addition additive in 3wt.%NaCl solution at room temperature are shown in Figure 2. The Nyquist plots of Ni plating in NaCl solution exhibit one capacitive loop in the entire frequency range. For measuring the corrosion behavior of the Ni deposits in NaCl solution, an equivalent circuit model consists of solution resistance (R_s), charge transfer resistance (R_{ct}) and constant phase element (CPE) has been proposed to simulate the metal/solution interface. The equivalent-circuit model for the corrosion behavior of Ni plating in the NaCl solution is shown in Figure 4.

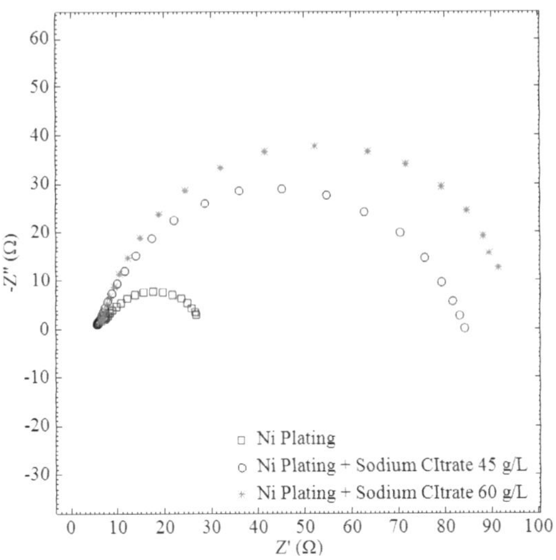

Figure 3. Electrochemical Impedance Spectra of SPCC Steel with electrodeposited nickel in the addition and without the addition of sodium citrate as additive

Table 3. Equivalent circuit parameters determined by modelling impedance spectra of Ni plating in 3 wt.% NaCl solution

	Sample	R_s ($\Omega.cm^2$)	R_{ct} ($\Omega.cm^2$)	Constant Phase Element ($\mu F.cm^{-2}$)
a.	Ni Plating	6.05	22.27	12.97
b.	Ni Plating + Sodium Citrate 45 g/L	5.32	78.7	53.22
c.	Ni Plating + Sodium Citrate 60 g/L	6.48	89.85	24.03

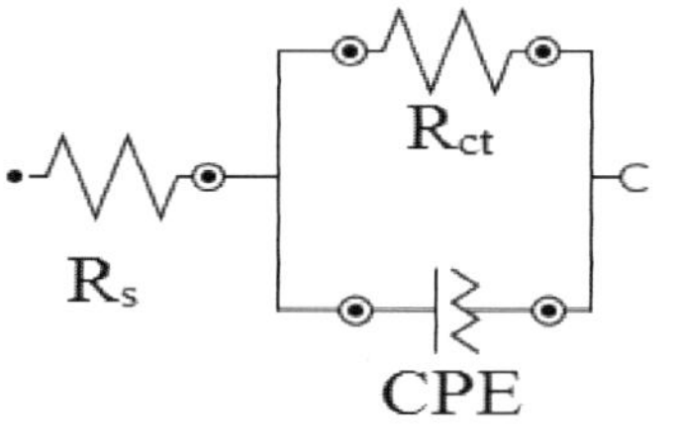

Figure 4. The equivalent-circuit model for the corrosion behavior of Ni plating in the 3 wt.% NaCl solution.

The equivalent circuit model is based on the following equations:

$$Z(\omega) = Rs + Zs/e(\omega) \quad (2)$$
$$Z_{s/e}(\omega) = 1/R_{ct} + 1/Y_0(j\omega)^{\alpha} \quad (3)$$

where ω, R_s, $Zs/e(\omega)$, R_{ct}, $Y_0$are the angular frequency, solution resistance, impedance of solution/ electrode interface, charge-transfer resistance and double layer capacitance, respectively.
Charge transfer resistance (R_{ct}) is the resistance offered by the metal atoms to get in contact with the electrolyte ionization time. The largest R_{ct} values have the highest corrosion resistance. Based on this research, the highest R_{ct} was owned by nickel plating with addition sodium citrate as much as 60g/L. The constant phase element (CPE) is used as a substitute for double-layer capacitor to fit the impedance data of the interface between an electrode and the surrounding electrolyte [19]. The simplified equivalent circuits shown in the insets of Figure 4 are used to model the electrode/electrolyte interface and the simulated results are shown in Table 3. The Nyquist plots (Figure 3) of the Ni plating in NaCl solution exhibit an increasing semicircle with decreasing grain size (addition sodium citrate 60g/L) and the corresponding R_{ct} values also increase. R_{ct} value of addition sodium citrate 60g/L is 4 times larger than that of Ni plating without addition additive. So, the refining grain size within addition sodium citrate as much as 60 g/L should exhibit better corrosion resistance in NaCl solution.
The increased corrosion resistance in nickel electroplating results is indicating due to the presence of passive layers. Nickel is one of the metals that easily form a passive layer. The electroplating process within the addition of sodium citrate will refining the grain size of the plating results where the grain has a high grain boundary density. On the formation of a passive layer in the presence of high grain boundary density will increase the number of active sites to form rapidly passive and protective passive film [20]. The layer of nickel plating has a lower grain size will have a high fraction of the passive layer due to the high density of nucleation site of the passive layer, so that the rate of corrosion that occurs in this layer will be lower [19].

4. Conclusions

[1]. The addition of sodium citrate as an additive will refine the grain size of nickel electroplating. The soluble sodium citrate additive in the electrolyte solution will be adsorbed and becomes a barrier from the diffusion path on the active surface of the grain growth. Impaired grain growth and increase in the frequency of the grain nuclei continuously will produce grains with finer size.

[2]. The corrosion resistance of the nickel plating layer with the addition of sodium citrate additive has increased. This can be seen from the value of corrosion rate that decreased.

[3]. The finer grain size produced by nickel electroplating with addition sodium citrate will have higher grain boundary fractions. This will lead to high nucleation density for the site of the passive layer, so that the passive layer is formed will be more stable. This stable passive layer will make the movement of electrons obstructed and the electrons are difficult to move to the surface to perform electrochemical reactions. The inhibited electrochemical reaction will decrease the corrosion rate and in other words increasing the corrosion resistance of the material.

[4]. Addition of 60 g/L sodium citrate has better impact to refinement grain size than 45 g/L. It also provide higher corrosion resistance on the perfomance of nickel electroplated.

5. References

[1]. C. Lin, J. Lee, H. Sheu, and S. Tsai, "Corrosion Resistance and Mechanical Property Enhancement of SPCC Steel Using an Induction Heat Treatment", International Journal of Electrochemical Science, vol. 12, pp. 7965 – 7976, (2017).
[2]. G. D. Bari, "Nickel Plating", in ASM Handbook, Volume 5, Surface Engineering (ASM International, USA, 1994), pp 742-765.
[3]. J. Mazia," Electroplated Coatings",in ASM Handbook, Volume 13, Corrosion (ASM International, USA, 1987), pp 942-975.
[4]. G. A. D. Bari, "Electrodeposition of Nickel", in Modern Electroplating Fifth Edition, edited by M. Schlesinger (John Wiley & Sons Inc, Hoboken, 2010), pp. 79–114.
[5]. T. Mooney, "Electroplated Coatings", in ASM Handbook , Volume 13A, Corrosion: Fundamentals, Testing, and Protection (ASM International, USA, 2003), 772–785.
[6]. W. Cheng, W. Ge, Q. Yang, and X. Qu, "Study on the corrosion properties of nanocrystalline nickel electrodeposited by reverse pulse current", Applied Surface Science, vol. (2013) , pp. 604– 608, (2013).
[7]. Q. Li-yuan, L. Jian-she, and J. Qing, "Effect of Grain Size on Corrosion Behavior of Electrodeposited Bulk Nanocrystalline Ni", Transaction of Nonferrous Metal Society of China, vol.20, pp.82-89, (2010).
[8]. L. Wang, J. Zhang, Y. Gao, Q. Xue, L. Hua, and T. Xua, "Grain Size Effect in Corrosion Behavior of Electrodeposited Nanocrystalline Ni Coatings in Alkaline Solution", Scripta Materialia, vol. 55, pp. 657–660, (2006).
[9]. N. P. Wasekar, P. Haridoss, S. K. Seshadri, and G. Sundararajan, "Influence of Mode of Electrodeposition, Current Density, and Saccharin on The Microstructure and Hardness of Electrodeposited Nanocrystalline Nickel Coatings", Surface & Coatings Technology, vol. 291, pp. 130–140, (2016).
[10]. ASTM B 832–93, Standard Guide for Electroforming with Nickel and Copper, American Society for Testing and Materials Annual Book, vol. 02.05.
[11]. U. Erb, A. M. El-Sherik, C. K. S. Cheung, and M. J. Aus, WIPO Patent No. 1994012695 A1 (9 June 1994).
[12]. B. L. Buckalew and T. A. Ponnuswamy, U.S. Patent No. 20150299882 A1 (22 October 2015).

[13]. R. Mishra and R. Balasubramaniam, "Effect of Nanocrystalline Grain Size on The Electrochemical and Corrosion Behavior of Nickel", Corrosion Science, vol. 46, pp. 3019–3029, (2004).
[14]. M. Stern and A.L. Geary, "Electrochemical Polarization I. A Theoretical Analysis of The Shape of Polarization Curves", Journal of The Electrochemical Society, Vol. 104, 56-63 (1957).
[15]. Kh. M. S. Youssef, C. C. Koch, and P. S. Fedkiw, "Improved Corrosion Behavior of Nanocrystalline Zinc Produced by Pulse-current Electrodeposition", Corrosion Science, vol. 46, pp. 51-64, (2014).
[16]. S. Wang, "Electrochemical Properties of Nano crystalline Nickel and Nickel-Molybdenum Alloys," Ph.D. Thesis, Queen's University, 1997.
[17]. A. M. Rashidi, and A. Amadeh, "Effect of Electroplating Parameters on Microstructure of Nanocrystalline Nickel Coatings", Journal of Materials Science & Technology, vol. 26,pp. 82-86, (2010).
[18]. F. Herdino, "Pengaruh Konsentrasi Aditif Sodium Sitrat dan Rapat Arus terhadap Ukuran Butir Lapis Listrik Nikel pada Baja SPCC," Thesis, Universitas Indonesia, 2017.
[19]. L. Qin, J. Lian and Q. Jiang, "Effect of Grain Size on Corrosion Behavior of Electrodeposited Bulk Nanocrystalline Ni", Transactions of Nonferrous Metals Society Of China, vol. 20, no. 1, pp. 82-89, (2010).
[20]. A. Balyanov, J. Kutnyakova, N.A. Amirkhanova, V.V. Stolyarov, R. Z. Valiev, X.Z. Liao, Y.H. Zhao, Y.B. Jiang, H.F. Xu, T.C. Lowe, and Y.T. Zhu, "Corrosion resistance of ultra fine-grained Ti". Scripta Materialia, vol. 51, pp. 225-229, (2004).

Acknowledgenments
The authors would say thanks to PT. Utama Raya Motor Industry and PITTA UI for all facilities and their financial support that offered to perform this work.

Optimization of palm oil extraction from Decanter cake of small crude palm oil mill by aqueous surfactant solution using RSM

Shewa Ahmadi Pirshahid[1,a] Wallop Arirob[2] and Vittaya Punsuvon[1,3,4,b]

1 Department of Chemistry, Faculty of Science, Kesetsart University, Bangkok 10900, Thailand
2 Department of Botany, Faculty of Science, Kesetsart University, Bangkok 10900, Thailand
3 Center of Excellence-Oil Palm, Kasetsart University, Bangkok 10900, Thailand
4 Center for Advance Studies in Tropical Natural Resource, NRU-KU, Kasetsart University, Bangkok 10900, Thailand
E-mail: [a]shewaa.ah@gmail.com, [b]fscivit@ku.ac.th

Abstract. The use of hexane to extract vegetable oil from oilseeds or seed cake is of growing concern due to its environmental impact such as its smelling and toxicity. In our method, used Response Surface Methodology (RSM) was applied to study the optimum condition of decanter cake obtained from small crude palm oil with aqueous surfactant solution. For the first time, we provide an optimum condition of preliminary study with decanter cake extraction to obtain the maximum of oil yield. The result from preliminary was further used in RSM study by using Central Composite Design (CCD) that consisted of thirty experiments. The effect of four independent variables: the concentration of Sodium Dodecyl Sulfate (SDS) as surfactant, temperature, the ratio by weight to volume of cake to surfactant solution and the amount of sodium chloride (NaCl) on dependent variables are studied. Data were analyzed using Design-Expert 8 software. The results showed that the optimum condition of decanter cake extraction were 0.016M of SDS solution concentration, 73°C of extraction temperature, 1:10 (g:ml) of the ratio of decanter cake to SDS solution and 2% (w/w) of NaCl amount. This condition gave 77.05% (w/w) oil yield. The chemical properties of the extracted palm oil from this aqueous surfactant extraction are further investigated compared with the hexane extraction. The obtained result showed that all properties of both extractions were nearly the same.

1. Introduction

One small crude palm oil mill located in the southern part of Thailand name Sangarun Palm Oil Co.,Ltd. (Thailand) produced crude palm oil around three tons per day. This mill not only produced palm oil but also by-products such as decanter cake, cake before filtration and cake after filtration were also produced too. All by-products had around 900-1,200 kilograms per day. The mill owner tried to make value added from the by-products by extraction the residue oil that left in all by-products. One popular technique is to use hexane as solvent for oil extraction.

Hexane has been broadly used in the vegetable oil industry due to its high oil extraction efficiency, over 99%. Vegetable oils are typically produced from oilseeds by either hexane extraction or a combination of mechanical processing and hexane extraction. However, there are growing health concerns and increased environmental regulations regarding the use of hexane in vegetable oil extraction. Exposure to hexane at 15 ppm/day for three months has been shown to cause peripheral nerve damage, and hexane is also a potential hazardous explosive material. Thus, extraction systems must fully contain hexane and provide for a leak detection. Future laws may be even more stringent; therefore, hexane-free extraction approaches are needed [1]. So, a new oil extraction technique is needed to replace hexane solvent. It is found that aqueous surfactant solution combined with sodium chloride technique is promising technique to replace hexane technique.

An aqueous extraction process, used water as a solvent, is one of the method to extraction vegetable oil. Currently, this method can achieve >70% of oil extraction efficiency [2]. However, to accomplish this yield, the extractor must execute at 50°C and a large amount of chemical reagent applied to adjust a pH to 12 [3].

A surfactant is an amphiphilic compound comprised of polar and non-polar parts, thus enabling the surfactant to reduce interfacial tension (IFT) of the system. By definition, IFT is the repulsive energy between two different species of molecules [4]. In other words, it is minimum energy required to separate one phase into another phase, increasing the surface area among the two phase. A high IFT value implies that the two species of molecules are very different in polarity [1]. The IFT value between water and vegetable oil is very high, between 20 and 30 mN/m [6]. Moreover, reduced IFT aids disruption of the oil trapped inside the oilseed matrix; the oil droplets are then readily transported to the aqueous phase [7]. Several reports showed that the IFT should be reduced to 0.01 mN/m (ultralow level) by incorporation of surfactant with co-solvent or use of an extended surfactant [5, 8-10].

Response surface methodology (RSM) is a useful statistical technique which has been applied in research into complex variable processes. It employs multiple regression and correlation analyses as tools to assess the effects of two or more independent factors on the dependent variables. Its principal advantage is the reduced number of experimental runs required to generate sufficient information for a statistically acceptable result. Response surface methodology has been successfully applied in the study and optimization of biodiesel production with rapeseed oil, soybean oil, *et cetera* [11].

Hence, the objectives of this work are: (1) maximize the oil recovery by the effect of temperature, surfactant concentration, ratio of cake to SDS solution and %NaCl (w/w) in decanter cake by using aqueous surfactant solution combined with sodium chloride; (2) determine the above optimum condition in decanter cake extracted using aqueous surfactant solution combined with sodium chloride by response surface methodology (RSM); (3) compare the properties of the crude palm oil obtained from hexane and SDS surfactant extraction.

2. Materials and Methods

Sodium dodecyl sulfate (SDS, biotechnology grade), Methanol (≥99.9% purity), hexane (99% purity). Only decanter cake was used in this research that obtain from small crude palm oil mill named Sangarun Palm Oil Co., Ltd. (Thailand).

Crude palm (4 g) was mixed with 40 mL of 0.02 M SDS solution. The mixture was mixed for 30 min, 70°C for temperature at 1,500 rpm (IKA, C-MAG HS7, Germany) on a digital hot plate/stirred and then centrifuged for 30 min at 8,000 rpm (CEN-D Nanasiam Intertrade, LDJ-5C). After centrifugation, two layers were obtained: liquid layer and solid layer. The liquid layer occurred on the top was separated and further heated. After that hexane was added into the liquid for oil recovery. The extracted oil was evaporated under vacuum evaporator to obtain crude palm oil. The palm oil extraction yield was calculated followed the Equation 1.

$$\text{Oil yield (\%)} = \left(\frac{\text{weight of oil}}{\text{weight of sample}}\right) \text{x } 100 \qquad (1)$$

Method

2.1 Preliminary study of oil extraction

Decanter cake was extracted by aqueous surfactant solution. There are four variable effects for study on oil extraction. These are concentration of sodium dodecyl sulfate (SDS) (0.01 M, 0.02M, 0.03M, 0.04M and 0.05 M), temperature (50°C, 60°C, 70°C, 80°C and 90°C), ratio of amount of cake to SDS (1g:5mL, 1g:10mL, 1g:15mL and 1g:20mL) and amount of sodium chloride (NaCl) (1%, 2%, 3%, 4% and 5% (w/w)) with keeping the constant of stirring at 1500 rpm and 30 min of contact time. The preliminary study is conducted for preliminary optimum condition that they will applied on RSM study.

2.2 RSM study of oil extraction

The obtained preliminary optimum condition was used to design the experiment for RSM and central composite design (CCD) using 5-level-4-factors. At five levels of independent variables from -2 to +2, 30 experiments were required. Surfactant concentration (A), temperature (B), the ratio of cake to SDS solution (C) and %NaCl (w/w) (D). The central values (zero level) chosen from preliminary studied for experimental design. The level and independent variable used for reaction were shown in Table 1. The experimental data were analyzed using a second order polynomial equation (Eq.2) to find the interaction between %oil yield and the independent variables.

$$Y = \beta_0 + \sum_{i=1}^{3} \beta_i X_i + \sum_{i=1}^{3} \beta_{ii} X_i^2 + \sum_{i=1}^{2} \sum_{j=i+1}^{3} \beta_{ij} X_i X_j \qquad (2)$$

where, Y is the respond (**%**oil yield), β_0 is the intercept, β_i, β_{ii} and β_{ij} are the linear, quadratic and interactive coefficients, respectively and X_i and X_j are the independent variables. Statistical analysis of equation was employed to evaluate the analysis of variance (ANOVA) and Design Expert 8 software (State Ease Inc., Minneapolis, Mn, USA) was used to design the experiments and carry out the regression analysis and plotting graphical of data. The independent variable and level used for reaction experiment parameter were designed as shown in table 1.

Table 1. Independent variables and levels for CCD of RSM

Variables	Symbol coded	Levels				
		- 2	-1	0	1	2
Surfactant concentration (M)	A	0.005	0.01	0.015	0.02	0.025
Temperature (°C)	B	50	60	70	80	90
Cake to SDS solution ratio (g:mL)	C	1:6	1:8	1:10	1:12	1:14
NaCl (%w/w)	D	1	1.5	2	2.5	3

2.3 Analysis of crude palm oil

Both crude palm oils obtained from hexane extraction using soxhlet extractor and SDS solution extraction were analyzed their properties for comparison. These properties were free fatty acid (FFA), iodine value (IV), deterioration of bleachability index (DOBI), saponification value (SV), Moisture content (MC) and Insoluble impurities.

3. Results and Discussion

3.1 The result from preliminary study of oil extraction

3.1.1 Effect of surfactant concentration

Increasing the SDS concentration from 0.01 M to 0.02 M resulted in increasing oil yield after that the oil yield decreased from the SDS concentration from 0.02 M to 0.05 M. At the critical micelle concentration (CMC) of each surfactant system, where the first micelle is formed, the IFT of the system is dramatically reduced [4]. In water, the CMC of SDS was 0.008 M at 25°C [12]**.** This reduces the IFT among the oil and water phases leading to degradation of oil into the surfactant solution [8, 9]. It can be that the CMC of SDS in the process of extraction should be higher than 0.008 M at 25°C. So, in this effect, the concentration of SDS at 0.02 M was selected for optimum condition.

3.1.2 Effect of Temperature

Increasing temperature from 50°C to 70°C resulted in increased oil yield after that the oil yield decreased from 70°C to 90°C. There was no increase in oil yield after 70°C (figure 2). One of explanation may be a reduction of IFT during the extraction temperature. The highest oil yield achieved at 70°C. This result could be explained by the IFT of the process. The IFT between palm oil and SDS solution on this work was small. Therefore, SDS solution reduced the heat energy required to disperse palm oil into the extraction medium. These results suggest to use of 0.02 M SDS at 70°C.

IOP Conf. Series: Materials Science and Engineering **348** (2018) 012011 doi:10.1088/1757-899X/348/1/012011

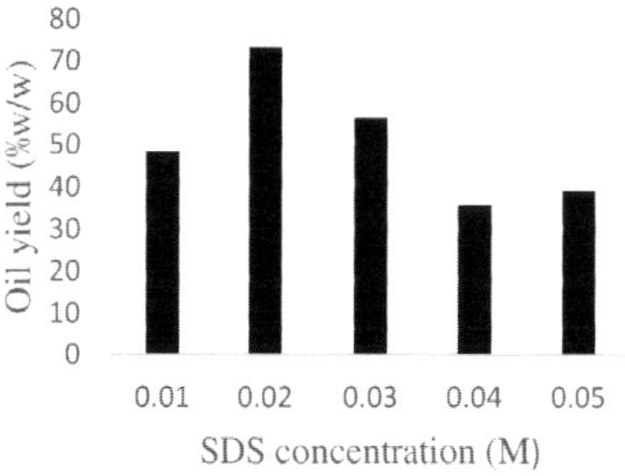

Figure 1 Effect of surfactant concentration on oil yield (conditions: 70°C for temperature with cake to solution 1g:10mL, stirring 1500 rpm for 30 min and 2%(w/w) of NaCl).

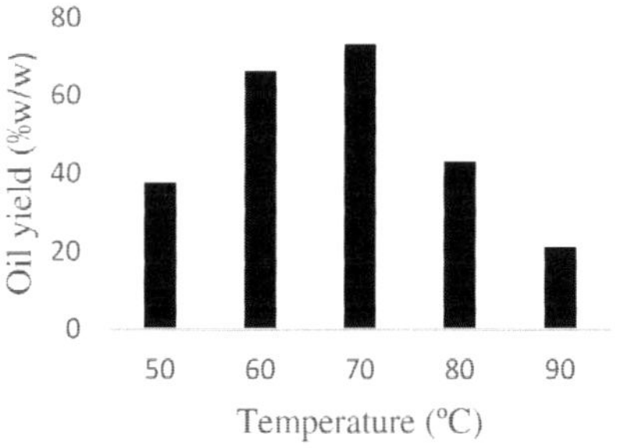

Figure 2 Effect of temperature on oil yield (conditions: 0.02 M SDS solution, cake to solution 1g:10mL, stirring 1500 rpm for 30 min and 2%(w/w) of NaCl).

3.1.3 Effect of Cake to SDS solution ratio

Experiments were conducted with 4 g of decanter cake at different amounts of cake to SDS solution. The ratio of cake to SDS solution from 1g:5mL to 1g:20mL had impact on the oil yield (figure 3). However, increasing the cake to SDS solution ratio to higher than 1g:10mL, oil yield was decreased. This may be similar to the "Effect of Temperature" that an increase in SDS concentration from 0.02 M to 0.03M resulted was dissimilarity. As explain above, increased SDS concentration in the system may increase emulsion formation.

3.1.4 Effect of %NaCl (w/w)

In recent studies, surfactants have been used for the extraction of oil from cruciferous [13] and palm kernel oil seeds by selecting optimum surfactant and salt concentrations based on interfacial tension (IFT) and phase behavior data [9]. The optimum salinity of the system denotes the electrolyte concentration where the surfactant system attained minimum IFT. Increasing the amount of NaCl from 1% to 2% resulted oil yield increase, after that the oil yield decrease from the amount of NaCl 2% to 5%. The result was shown in figure 4. The obtained highest oil yield with NaCl is at 2% (w/w).

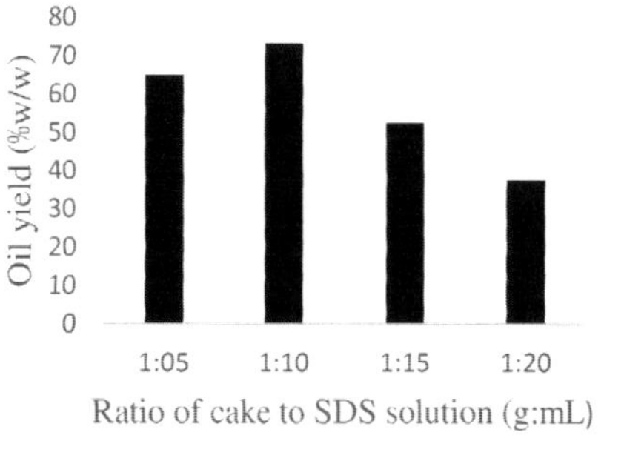

Figure 3 Effect of cake to SDS solution on oil yield (conditions: 0.02 M SDS solution, stirring 1500 rpm for 30 min at 70°C and 2%(w/w) of NaCl).

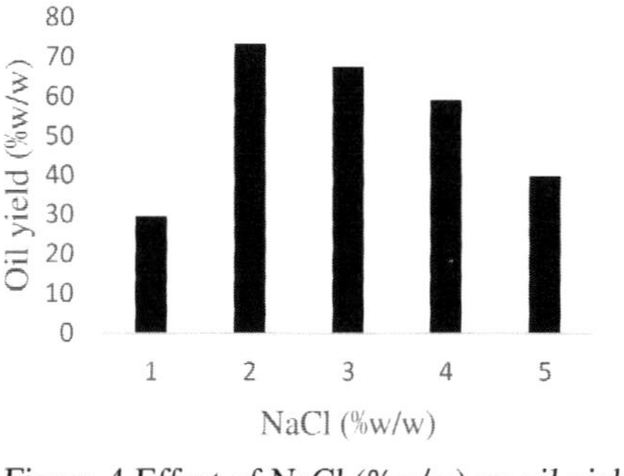

Figure 4 Effect of NaCl (%w/w) on oil yield (conditions: 0.02 M SDS solution, stirring 1500 rpm for 30 min at 70°C and cake to solution 1g:10mL).

The summary results from preliminary study of oil extraction from figure 1 to figure 4 were 0.02 M of SDS concentration, 70°C, 1g:10mL of cake to SDS ratio and 2% of NaCl amount. These conditions were used in CCD of RSM study.

3.2 The results from RSM study of oil extraction

The conditions for aqueous extraction of oil from decanter cake were optimized by RSM using CCD in table 2 shows the experimental design with code variable observed and predicted values of %oil yield. A Design-Expert 8 software was employed to determine and evaluate the coefficients of the full regression model equation and their statistical significance. The second polynomial model for the %oil yield is regressed as shown in Equation 3

$$Y = 76.39+3.25A+4.38B+3.94C+2.66D-3.69AB-0.26AC-1.63AD - 0.2BC+0.75BD+1.08CD-1.43A^2-3.53B^2-3.97C^2-4.64D^2 \quad (3)$$

Where A, B, C and D correspond to the coded values of the four variables (surfactant concentration, temperature, ratio of cake to surfactant and %NaCl). The oil yield was in the range of 42.91-73.93% (w/w). The p value of the model was less than 0.0001 indicating that the model was significant. In addition, the suitability of model was also tested by using the regression equation determination coefficient (R^2). Regression analysis showed the coefficient of determination ($R^2 = 0.9002$) was satisfactory for validating the significant of the model. All the independent variables (A, B, C and D), four quadratic terms (A^2, B^2, C^2 and D^2) and three interaction terms (AB, BC and BD) had significant effect on Y (p<0.05). In figure 5 and figure 6 shown the results express the interactions between the two independent variables while the another two variables were maintained at the central point.

Table 2. Experimental design with observed and predicted values of crude palm oil

Run	Surfactant concentration (M)	Temperature (°C)	Cake to surfactant ratio (g:mL)	NaCl (%w/w)	Observed Oil (%)	Predicted Oil (%)
1	-1	-1	-1	-1	42.91	45.74
2	-1	-1	-1	+1	44.53	41.73
3	-1	-1	+1	-1	58.56	56.73
4	-1	-1	+1	+1	65.11	63.95
5	-1	+1	-1	-1	48.82	50.43
6	-1	+1	-1	+1	55.53	52.88
7	-1	+1	+1	-1	63.61	60.94
8	-1	+1	+1	+1	72.02	73.49
9	+1	-1	-1	-1	61.92	60.43
10	+1	-1	-1	+1	58.70	56.78
11	+1	-1	+1	-1	63.78	62.45
12	+1	-1	+1	+1	64.06	65.43
13	+1	+1	-1	-1	66.56	67.80
14	+1	+1	-1	+1	67.32	65.42
15	+1	+1	+1	-1	67.98	66.97
16	+1	+1	+1	+1	70.91	67.43
17	-2	0	0	0	45.73	43.56
18	+2	0	0	0	66.82	65.03
19	0	-2	0	0	53.72	52.45
20	0	+2	0	0	67.13	66.03
21	0	0	-2	0	49.08	51.45
22	0	0	+2	0	68.35	70.34
23	0	0	0	-2	68.19	70.94
24	0	0	0	+2	71.56	70.34
25	0	0	0	0	73.65	75.76
26	0	0	0	0	72.86	75.76
27	0	0	0	0	72.94	75.76
28	0	0	0	0	73.11	75.76
29	0	0	0	0	73.93	75.76
30	0	0	0	0	73.46	75.76

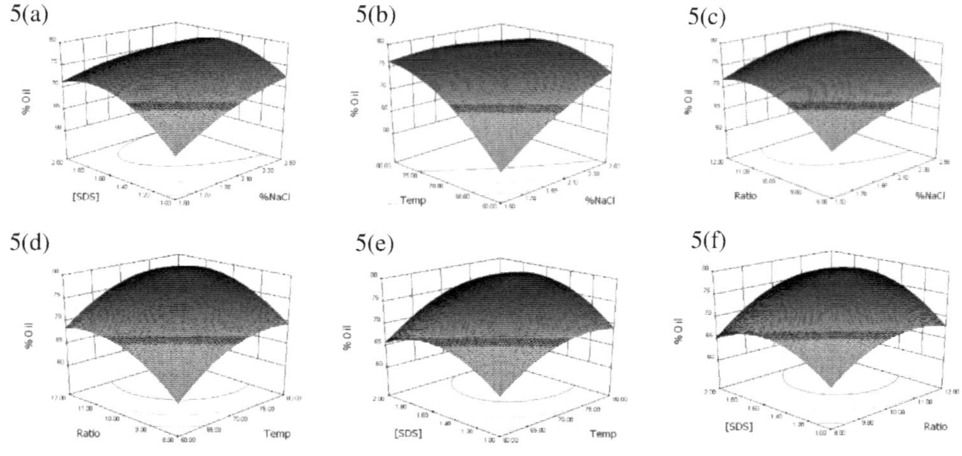

Figure 5. The 3-D response surface plot for the effect of (a) %NaCl/SDS concentration, (b) %NaCl/temperature, (c) %NaCl/ cake to SDS solution ratio, (d) temperature/cake to SDS solution ratio, (e) temperature/SDS concentration and (f) cake to SDS solution ratio/SDS concentration on oil yield.

The response surface plot of %oil yield from reaction with independent variables was shown in figure 5. The response surface plot showed the optimal condition between the various variables. Figure 5a, 5b and 5c showed the interaction between %NaCl (D) with SDS concentration (A), temperature (B) and cake to SDS solution ratio (C) respectively. It revealed the increased %oil conversion with increasing the variables. In addition, the figure 5d and 5e showed the interaction between temperature (B) with SDS concentration (A) and cake to SDS solution ratio (C). The figure 5f showed the interaction between cake to SDS solution ratio (C) with SDS concentration (A). The %oil yield increased with increasing of variables.

The optimum condition of this process was predicted by applying numerical optimization of Design-Expert 8 software using RSM. The oil production was produced with high oil yield (77.05%) at 0.016 M of SDS at 73°C of temperature, 1g:10mL of cake to SDS solution ratio and 2% of NaCl (w/w) amount by constant keeping 1,500 rpm of stirring rate and 30 min of extraction time.

3.3 The result of oil analysis

The quality of the difference extraction crude palm oil was tested by following the standards of AOCS. The results were shown in Table 3.

Table 3. Properties of extraction palm oil compared between surfactant and hexane

Parameter	Test Method	Extracted by SDS	Extracted by hexane	Limit crude palm oil
Free Fatty acid (%w)	AOCS official method Ca 5a-40	7.92	7.75	$\leq$ 5.00
Iodine value (g I/100 g)	AOCS official method Cd 1-25	47.71	48.28	$\geq$ 50.00
DOBI	Developed by Dr. P.A.T. Swoboda	0.95	0.97	$\geq$ 2.00
Saponification (mg KOH/g)	-	180.76	182.27	-
Moisture content (%w)	-	0.71	0.69	$\leq$ 0.50
Insoluble impurities (%w)	-	0.0055	0.0037	$\leq$ 0.50
Oil yield (%w/w)	-	77.05	78.36	-

4. Conclusion

The aqueous surfactant extraction offered several advantages when compared with hexane extraction. The aqueous surfactant extraction method can be operated at ambient temperature using non-toxic and nonflammable solvents. Surfactant concentration, temperature, ratio of cake to SDS solution and %NaCl (w/w) were impacted on oil yield. The highest yield oil extraction after using RSM was 77.05% respectively, using SDS 0.016 M at 73°C with cake to SDS solution ratio 1g:10mL, stirring at 1,500 rpm for 30 min and 2% NaCl (w/w). The properties of decanter cake were compare with extracted by aqueous surfactant and Soxhlet hexane extraction, the results shown that all properties of both extractions were nearly the same.

Acknowledgements

This work was supported by the Thailand research fund (TRF) under the research and researcher for Industry: RRI (MSD59I0061) and Sangarun Palm Oil Co.,Ltd. (Thailand). The author also thank Center of Excellence-Oil Palm Kasetsart University, Center for Advance Studies in Tropical Natural Resource (NRU-KU) and the Graduate school and Faculty of Science, Kasetsart University for partial supported.

References

[1] Tuntiwiwattanapun N, Tongcumpou C, Haagenson D and Wiesenborn D 2013 *J. Am. Oil Chem. Soc.* **90** 1089-99

[2] Rosenthal A, Pyle D and Niranjan K 1996 *Emzyme and Microb. Technol.* **19** 402-20

[3] Rosenthal A, Pyle D and Niranjan K 1998 *Trans. Inst. Chem. Eng.* **76** 224-30

[4] Rosen M 2004 Surfactant and interfacial phenomena, Canada

[5] Do L and Sabatini D 2010 *J. Am. Oil Chem. Soc.* **87** 1211-20

[6] Gunstone F 2009 Rapeseed and Canola Oil: Production, Processing, Properties and Uses, USA: Blackwell

[7] Campbell K and Glatz C 2009 *J. Agric. Food Chem.* **57** 10904-12

[8] Kadioglu S, Phan T and Sabati A 2011 *J. Am. Oil Chem. Soc.* **88** 863-69

[9] Naksuk A, Sabatini D and Tongcumpou C 2009 *Ind. Crops. Prod.* **30** 194-98

[10] L. Do and D. Sabatini D 2011 *Sep. Sci. Technol.* **46** 978-85

[11] G. Vicente, A. Coteron, M. Martinez and J. Aracil 1998 *Ind. Crops. Prod.* **8** 29-35

[12] [Online]. Available: http://wolfson.huji.ac.il/purification/PDF/detergents/PIERCE_DetergentRemoval.pdf. [Accessed 14 May 2012].

[13] L. Ugolini, G. Nicola and S. Palmieri 2008 *J. Agric. Food Chem.* **56** 1595-1601

IOP Conf. Series: Materials Science and Engineering **348** (2018) 012012 doi:10.1088/1757-899X/348/1/012012

Effect of silica-palm shell carbon composite additive in enhancing the strength of the concrete in the oil-well cementing job

A Novriansyah[1], U Mursyidah[1], Novrianti[1], S S Putri[1], and S S Riswati[2]
1 Department of Petroleum Engineering, Universitas Islam Riau, Pekanbaru, Riau, Indonesia
2 Department of Petroleum Engineering, Universitas Trisakti, Jakarta, Indonesia

Abstract. This study provides an analysis of composite additive effect to concrete's strength in the oil-well cementing job. The composite additive is originated from the nano-sized form of silica and charcoal from palm shell waste. The quality of the concrete will be determined from its porosity, compressive strength, and shear bond strength parameters. Those parameters must be reliable base on the most respectable standards in oil and gas industry, in this study we use the standard from American Petroleum Institute (API). Six concrete samples with different concentration will be tested to obtain these parameters. The result from the test shown a decrement trend of the porosity while the concentration is increased. In contrast, the highest values of compressive strength and shear bond strength are obtained from the sample with higher additive concentration. The optimum strength was obtained in sample with 0.02% The results become clearly proven through verification by scanning electron image where the additive has successfully fill the voids in the concrete's sample, resulting in strength enhancement of the sample.

1. Introduction

Application of Nanotechnology transformed every industrial sector, particularly in energy industry. A precise engineering process and matters control at the scale of nanometer creates a new perspective in developing advanced material [1]. Through this innovative technology application, a product of nanotechnologies such as nanoparticles, nanostructure, nanosensor, and nanobots can be produced. These products can be utilized as an additive for improving the properties of the object material [2]

In the renewable energy sector, utilization of solar and fuel cell have been developed further by adopting nanotechnology [3]. An optimist signs also arose from the nonrenewable energy area where nanomaterial is successfully implemented in laboratory scale for enhancing the crude oil recovery and maintaining the lifetime of concrete, which is critical in supporting casing in the wellbore[4].

Quality of concrete in the wellbore is one of vital consideration to prolong the lifetime of oilwell structure. A longer lifetime of the well structure will reduce the cost of concrete quality improvement due to the bad bonding of cement concrete, causing micro annulus and fracture which is lead to casing corrosion and will affect to well performance in transporting liquid and gas phase of hydrocarbon from the reservoir to the surface. There are three parameters describes the quality of the concrete in an oil well construction. First is porosity, a parameter which define as a ratio of voids volume to the total volume of the concrete. The others are Compressive Strength (CS) and Shear Bond Strength (SBS), parameters which indicate the strength of the concrete in resisting load or pressure from the reservoir or mechanical load due to production job in the well. Moreover, the concrete must hold the tendency

of crack generation in the casing[5]. Based on API Standards article 10 A, the minimum values of the CS and SBS are 500 psi and 100 psi respectively [6].
Progressive development of Nanotechnology become eminent over a decade in the concrete industry. Works by Li et al.(2004) have successfully increased the mechanical strength of the cement mortar through nano SiO_2 and nano Fe_2O_3[7]. Another study by Campillo et. Al.(2007) have utilized nanoalumina for modifying the strength of belite cement, which friendly environmental and efficient[8]. In oil well construction, a study regarding the application of nanoparticle additive has been performed by Patil and Desphande (2012). Through laboratory scale experiment, they conclude that addition of nanosilica can increase strength and reduce fluid loss of the concrete [4]. Although application concrete strength has shown positive results by adding nanomaterial, an actual field-scale implementation still far away due to the cost of the nanomaterial additive. Therefore, another additive which is less-expensive should be needed to create a concrete not only strong enough but also economically friendly.
Through our experimental study, we introduced a nanocomposite additive, a combination of nanosilica and Palm-Shell Charcoal (PSC). The reason of using PSC is based on the fact that the source is abundant in Indonesia, where the palm oil plantation area grows up to 10.7 million hectares in 2014, increasing 21.5 % during 4 years since 2010. Moreover, the production reached 29.7 million tonnes in 2014 or increase 25 % since 2010[9], resulting in a huge source of palm-shell waste. Combination of these additives and cement slurry will be tested to observe the effectivity of the additives to porosity and strength of the concrete. A positive result from this study can be a big expectation for transforming an almost noneconomical waste product into a useful product with a better economic value and as an effort keep the environment clean.

2. Methodology

The experimental study divided into two distinct experiments. First is porosity determination where the objective is to investigate the effect of PSC nanocomposite in filling the voids in the concrete sample. The other experiment is biaxial loading test for determining CS and SBS parameter of the sample. From this results, the effect of the nanocomposite in enhancing the concrete can be observed.

2.1 Materials

API-Class G cement will be used to make concrete samples because of the oil-well condition where high-pressure and high-temperature environment tends to contact with the concrete. Class G cement can be utilized up to 8000 ft depth of cementation[10]. Nanosilica is synthesized by using the sol-gel method from Tetra Etil Organo Silicate (TEOS) material. The size of silica particle which is formed by using this method is in the range of 10 to 20 nanometers (Table 1). Palm shell waste is acquired from the one of palm-oil factory in Indonesia. Palm shell will be burned at 500°C by using the electric oven. The charcoal Shells are ground and sieved to obtain a uniform sized of PSC. From its chemical composition in Table 2, PSC contains rich of silica or SiO_2 compound. Other material such as bentonite and calcium chloride will be added to the mixture to keep the slurry water content and to accelerate the time for the slurry to harden.

Table 1. Physical properties of nanosilica

Physical Properties	Value
Density	2.17 – 2.66 gr/cm^3
Melting Point	± 1700 °C
Boiling Point	2230 °C
Color	White
Particle Size	10 – 20 Nanometer
Bulk Density	0.011 gr/ml

Table 2. Chemical of composition of PSC

Chemical	wt%
MgO	4.9
Al_2O_3	2.2
SiO_2	30.1
P_2O_5	19.6
SO_3	3.28
K_2O	17.5
CaO	14.6
Fe_2O_3	5.08
CuO	0.388
FeO	2.30

2.2 *Procedure*

Various concrete samples were prepared by mixing cement powder, de-ionized water, nano silica, palm shell charcoal and other additives. The material composition is based on Table 3. The mixture is mixed by using a high-speed propelled mixer. The velocity is constantly maintained at 4000 RPM for 5 minutes to achieve a homogeneous mixture of the slurry sample. 2 types of concrete mold, cubic and cylindrical are prepared for determining porosity, Cs, and SBS parameters. The cubic sample will be used CS test while the Cylindrical sample is needed for porosity and SBS test. Each slurry is then transferred into the concrete mold and dried at 180°F for 24 hours to ensure that all of the concrete samples are completely harden.

Table 3. Composition of nanocomposite in cement slurry

Sample Name	Composition
A	0.02% Nanosilica
B	0.02% Nanosilica + 1% PSC
C	0.02% Nanosilica + 2% PSC
D	0.02% Nanosilica + 2.5% PSC
E	0.02% Nanosilica + 3% PSC
F	0.02% Nanosilica + 3.5% PSC

Each sample will be measured its porosity by implementing helium porosimeter apparatus. Samples are dried for 24 hours again in the oven at 180°F for re-ensuring no trapped liquid inside the sample. The bulk volume and mass of the Dried sampled will be calculated and weighted. Measured the dead volume of the apparatus by placing steel plug into the chamber and record the helium volume as dead volume data. Remove the steel plug from the chamber and placed the sample inside the chamber. Measured and recorded the volume of helium gas inside the chamber. The porosity can be determined by dividing the differential volume of helium gas when the sample inside the chamber minus dead volume with sample bulk volume.

The other experiment, CS parameter can be obtained from the biaxial loading test by using hydraulic test apparatus. The dried cubic sample is placed below the hydraulic piston before initiating the test. A Certain amount of pressure will be loaded to the piston, causing the piston moves downward and press the sample. The sample will withstand the vertical load from the piston, causing the increment of vertical pressure due to the loading will be continuously given. The pressure where the failure condition happened to the sample is recorded. CS is a ratio of recorded pressure to sample cross-sectional area. The test is then repeated for SBS determination by exchanging the cubic sample with the cylindrical sample. Scanning Electron Microscope (SEM) observation will be done to image the effect of nanocomposite on the concrete surface structure.

3. Result and discussion

The result from the porosity calculation on each sample shows an indication that PSC effectively fills the void of the concrete samples, causing a lower value of porosity. From Figure 1, the porosity of sample tends to reduce if the PSC concentration increase. Escalating the concentration up to 3 wt% resulting a decremental porosity to approximately 1%. Comparing the sample without PSC to another sample with 3 wt% of PSC, decremental value porosity reaches 93%. This porosity result is consistent with other experimental study by Chindaparist et. al (2005) where they utilized fly ashes material, which had larger particle size (6.4 micrometer), compare to nanosilica (20 nanomater)[11]. Image from the SEM observation also verifies the effectivity of PSC in filling the voids of the sample. Comparing Image sample A (Figure 2a) and sample E (Figure 2b) confirms that PSC has successfully sealed the pores, which is indicated by darker spot in the SEM image. Lower porosity will minimizes the possibility of fluids trapped inside, resulting a compact bonding of the concrete and higher strength value [4].

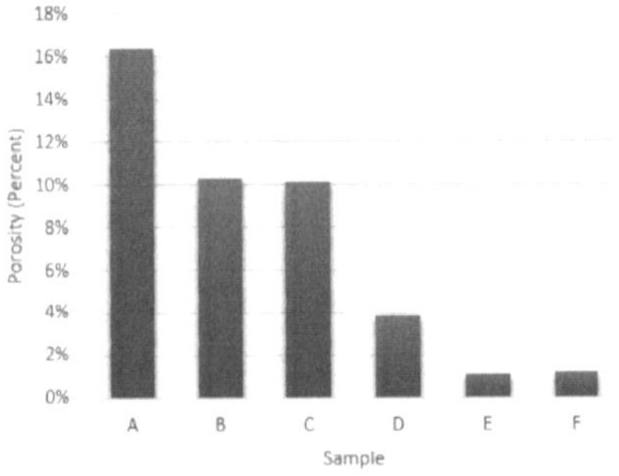

Figure 1. Porosity of each sample.

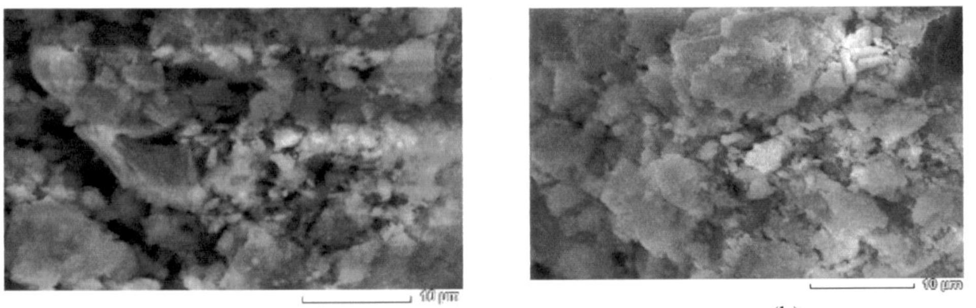

(a) (b)

Figure 2. SEM Image of Sample A (a) and sample E (b).

Effect of Nanosilica-PSC nanocomposite to CS and SBS of each sample are plotted in Figure 3a and 3b respectively. All of CS and SBS values of each sample is satisfy the minimum requirement of CS and SBS value based on API standards. These plots prove that PSC effectively enhances the strength of the concrete. At low concentration of PSC (sample B and C), The CS and SBS values do not give a significant incremental value, compare sample without PSC (Sample A). however, for a sample with high concentration of PSC (Sample E), the CS and SBS rise up to 1000 and 150 psi respectively, 42% and 54% incremental compared to sample A. result from the sample F illustrates the more additional PSC is not always resulting in a concrete with better strength, concludes that sample E reveals an optimum condition where porosity is the lowest and highest for CS and SBS.

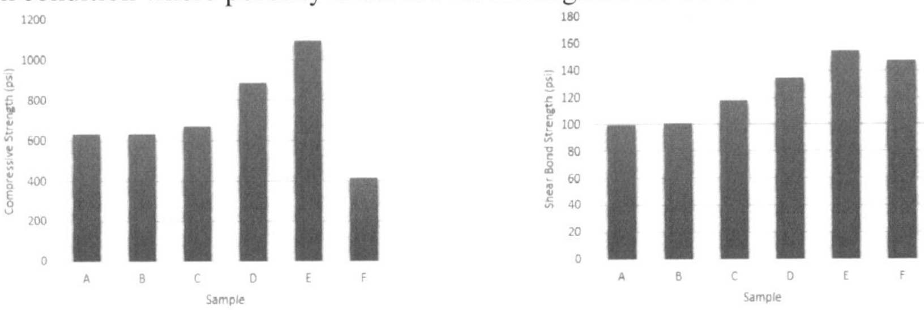

(a) (b)

Figure 3. Compressive Strength (a) and Shear Bond Strength (b) of each sample.

4. Conclusion

An experimental study has been performed to investigate the effect of PSC nanocomposite on porosity and strength of the concrete. Six samples are prepared and following a serial test for obtaining porosity, CS, and, SBS values. Results from test have shown that the porosity tend to decrease when the concentration of PSC is increased. This is confirmed by results from SEM analysis where more darker spot can be found in the sample without PSC, means more voids in concrete sample is filled by an additive.another result from the loading test yields a higher value of CS and SBS for samples with higher concentration of PSC. In this test, optimum value is obtained for sample with 0.02% nanosilica and 3% PSC.

5. References

[1] Denney, D. 2014. Nanotechnology applications for challenges in Egypt. *Journal of Petroleum Technology*, *66*(02), 123-126.

[2] Mokhatab, S., Fresky, M. A., & Islam, M. R. 2006. Applications of nanotechnology in oil and gas E&P. *Journal of petroleum technology*, *58*(04), 48-51.

[3] Mao, S. S., & Chen, X. 2007. Selected nanotechnologies for renewable energy applications. *International journal of energy research*, *31*(6 - 7), 619-636.

[4] Patil, R. C., & Deshpande, A. 2012. Use of nanomaterials in cementing applications. In *SPE International Oilfield Nanotechnology Conference and Exhibition*. Society of Petroleum Engineers.

[5] Goodwin, K. J., & Crook, R. J. 1992. Cement sheath stress failure. *SPE Drilling Engineering*, *7*(04), 291-296.

[6] Specification, A. P. I. 2002. 10A, Specification for cements and materials for well cementing. *American Petroleum Institute Publishing Services. Washington DC.*

[7] Li, H., Xiao, H. G., Yuan, J., & Ou, J. 2004. Microstructure of cement mortar with nano-particles. *Composites Part B: Engineering*, *35*(2), 185-189.

[8] Campillo, I., Guerrero, A., Dolado, J. S., Porro, A., Ibáñez, J. A., & Goñi, S. 2007. Improvement of initial mechanical strength by nanoalumina in belite cements. *Materials Letters*, *61*(8), 1889-1892.

[9] Indonesia, S. P. 2013. Indonesia 2012-2014: Kelapa Sawit. *Direktorat Jenderal Perkebunan, Kementerian Pertanian.*

[10] Nelson, E. B. (Ed.). 1990. *Well cementing* (Vol. 28). Newnes.

[11] Chindaprasirt, P., Jaturapitakkul, C., & Sinsiri, T. (2005). Effect of fly ash fineness on compressive strength and pore size of blended cement paste. *Cement and Concrete Composites*, *27*(4), 425-428

IOP Conf. Series: Materials Science and Engineering **348** (2018) 012013 doi:10.1088/1757-899X/348/1/012013

Numerical analysis of dynamic behavior of pre-stressed shape memory alloy concrete beam-column joints

S Yan[1], Z F Xiao[1], M Y Lin[1] and J Niu[1, 2]

1. School of Civil Engineering, Shenyang Jianzhu University, Shenyang Liaoning, 110 168, China.
2 Faculty of Infrastructure Engineering, Dalian University of Technology, Dalian Liaoning, 116024, China.
Corresponding author email: syan1962@163.com

Abstract. Beam-column joints are important parts of a main frame structure. Mechanical properties of beam-column joints have a great influence on dynamic performances of the frame structure. Shape memory alloy (SMA) as a new type of intelligent metal materials has wide applications in civil engineering. The paper aims at proposing a novel beam-column joint reinforced with pre-stressed SMA tendons to increase its dynamic performance. Based on the finite element analysis (FEA) software ABAQUS, a numerical simulation for 6 beam-column scaled models considering different SMA reinforcement ratios and pre-stress levels was performed, focusing on bearing capacities, energy-dissipation and self-centering capacities, etc. These models were numerically tested under a pseudo-static load on the beam end, companying a constant vertical compressive load on the top of the column. The numerical results show that the proposed SMA-reinforced joint has a significantly increased bearing capacity and a good self-centering capability after unloading even though the energy-dissipation capacity becomes smaller due the less residual deformation. The concept and mechanism of the novel joint can be used as an important reference for civil engineering applications.

1. Introduction

Reinforced concrete (RC) beam-column joints are dominant parts of a RC structure, which plays an important role in high-rise structures. Under a dynamic environment such as an earthquake, the well dynamic performance is important to enable to ensure the reliability of the structure. The main dynamic performances are including the bearing capacity, deformation capacity, energy-dissipation capacity, etc. One of the traditional design criteria is meeting requirements of enough bearing capacity and well deformation capacity to dissipate more energy and the structural stability during a dynamic loading. However, a traditional RC structure might have an observable residual deformation after unloading which could greatly influence the vertical stability of the structure even inducing severe damages or collapse. Self-centering capacity of a structure is a kind of new but important concept which represents a recovery capacity to its original or balance position by its own ability [1-3]. Smart materials are usually used to design the self-centering devices or resembled into structures to make them having the self-centering function [4].

Shape memory alloy (SMA) is a new type of smart metal material. SMA has obviously superior shape memory effect (SME) and superelasticity (SE) characteristics which has both sensing and driving functions [5]. The SE effect of SMA is a process of stress-induced martensitic transformation after loading of A_s-A_f, a process of austenite transformation after unloading, and the material is completely restored to the original shape. The SE type SMA has a unique hysteresis characteristic which can consume large amounts of energy in the material without causing residual strain or deformation. The SE of SMA can also effectively improve the bearing capacity, residual deformation and energy dissipation capacity of beam-column joints [6-8].

Based on the SE of SMA, a novel RC beam-column joint reinforced with SMA tendons for part instead of traditional reinforcement is proposed in the paper, aiming at improving the self-centering performance under dynamic loading. A finite element model is established by finite element analysis (FEA) software (ABAQUS) to analyze dynamic behaviors of the proposed novel joints.

2. Mechanical properties of SMA

In the SMA excellent properties, the most significant features are SME and SE. The SMA elastic strain is up to 5% -20% but the elastic strain for an ordinary metal material elastic strain is generally not more than 0.5% [9]. In a cycle of loading and unloading, the stress-strain curve of the superelastic SMA will form a complete hysteresis loop, thus providing good energy dissipation and self-centering capability. The SE of SMA is affected by alloy composition and heat treatment method, ambient temperature, strain rate, strain amplitude, cyclic loading, pre-strain and other factors. Huang [10] treated Ti-49.8at% Ni alloy wire with cold drawing and intermediate annealing process. The results show that the material can obtain complete nonlinear ultra-linearity after large cold deformation and medium temperature annealing elasticity. When the Ni content is more than 50%, the material obtain good superelasticity after heat treatment, such as Ti-50.8at% Ni alloy wire at 400℃ for 30 min after quenching to obtain better superelasticity. Strain amplitude has a great influence on the energy dissipation of materials [11-12]. With increase of the strain amplitude, the martensitic transformation of SMA is more obvious and the energy consumption increases, but the change of critical stress is small. The equivalent damping ratio increases linearly in the range of medium and low strain amplitude (1% -5%). Since the martensitic elastic deformation occurs after the normal phase transition, the equivalent damping ratio is higher than that at high strain (5% -10%). When the SMA performance is not stable, the residual strain increases with increase of the strain amplitude. Ip[13] predicted the energy dissipation of SMA wire under bending load, and the results show that pre-strain can improve the damping capacity of SMA. In the beam repair performance test, SMA tendons was loaded in the effective pre-strain with the greater strain, the fracture recovery rate was better [14].

3. Finite element model

3.1. Model description

The numerical simulation model is selected as a typical member of a multi-layer and high-rise frame RC structure under a horizontal load, and the model scale is 1:2. In this paper, two types of beam-column joints are designed. One is the ordinary beam-column concrete joint and the other is reinforced with pre-stressed SMA tendons. The geometrical dimensions of the beam members are 770mm × 120mm × 240mm (length × width × height), and the column member geometry is 1500mm × 200mm × 200mm (length × width × height). The ordinary beam is reinforced by 4 longitudinal reinforcement of HRB400Φ10, stirrup of HRB400Φ10 @ 80mm, and the protective layer thickness of 20 mm. The column is installed by 4 longitudinal reinforcements of HRB400Φ16, stirrup of HRB400Φ10 @ 80mm, and the protective layer thickness of 20 mm. Specific specimen cross-section size and reinforcement situation are shown in Figure 1. The specimen design parameters are shown in Table 1.

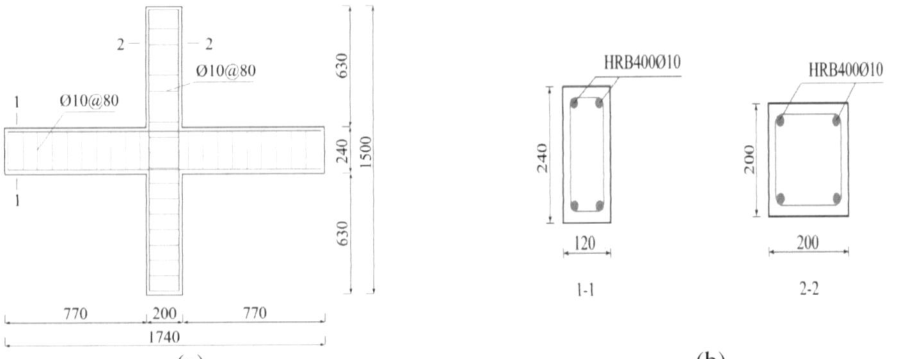

(a) (b)

Figure 1. The model size and reinforcement. (a) The model dimension and reinforcement. (b) Reinforcements in cross-sections.

Table 1. The model design parameters

Specimen number	Column reinforcement	Reinforcement of beam members		SMA tendons	
	Longitudinal tendons (mm)	Stirrups(mm)	Longitudinal tendons (mm)	Reinforcement ratio(%)	Pre-strain(%)
XJD-1	4Φ16	Φ10@80	4Φ10Ordinary tendons	0	0
YJD-1	4Φ16	Φ10@80	4Φ10 SMA tendons	0.65	4
YJD-2	4Φ16	Φ10@80	4Φ10 SMA tendons	0.65	5
YJD-3	4Φ16	Φ10@80	4Φ10 SMA tendons	0.65	6
YJD-4	4Φ16	Φ10@80	4Φ12 SMA tendons	0.94	6
YJD-5	4Φ16	Φ10@80	4Φ14 SMA tendons	1.19	6

3.2. Constitutive model of concrete

The constitutive model of CONTRETE02 [15] in OPSEES platform is used in simulation analysis. The compression skeleton curve is used by Kent and Park, and Scott [16] modified the Kent-Park model. The hysteresis rules of constitutive relation of concrete: from the unloading point D on the skeleton curve, the straight line DE with the slope of the initial tangential stiffness is unloaded to a certain extent and then unloaded to the residual plastic strain point H by a straight line EH with a slope of half of the reload slope, the reload line is a straight line connecting the plastic strain point H to the unloading point D and is loaded back along the skeleton line back to the skeleton line. Residual plastic strain depends on unloading strain and reloading stiffness. The slope of the reload is determined by defining the point R, where all the reload lines extend in the opposite direction. The slope of the reload line can be obtained from the unload point and the R point. That is as shown in Figure 2.

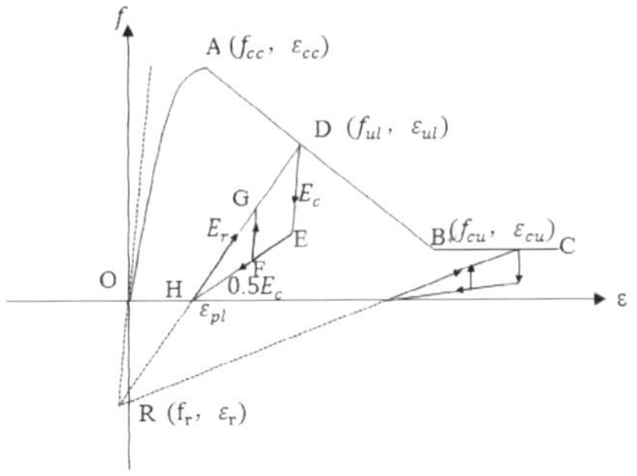

Figure 2. The hysteresis rules of constitutive relation of concrete.

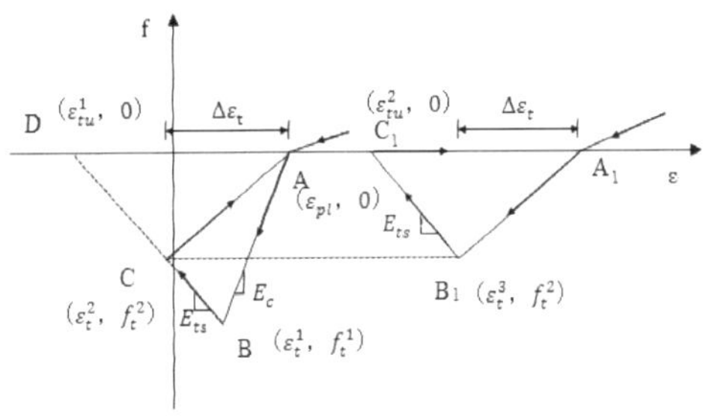

Figure 3. The tensile part of the concrete skeleton curve.

The tension part is proposed by Yassin [17]. The skeleton line is divided into two straight lines. The first one is linearly loaded to the tensile peak stress, and then along the straight line from the tensile peak stress point goes down to stress zero. The tension loading and unloading lines overlap, being pulled into the starting point of the linear plastic strain residual tension by unloading point. The model takes into account the stiffness degradation during the unloading process of concrete and the hysteretic energy dissipation of the loading and unloading process. That is shown in Figure 3.

3.3. Reinforcement constitutive model

The constitutive model of USTEEL02 [18] in PQ-fiber platform is used in the simulation analysis. Its main feature does not immediately point to the history of the largest point, but first by unloading stiffness loaded to the maximum point of the corresponding stress of 0.2 times, that is, $0.2f_{max}$, and then point to the largest point in history in the reverse reload. The specific process is shown in Figure 4.

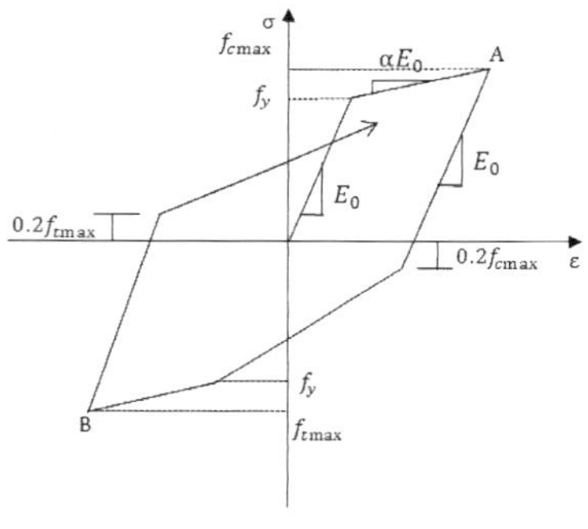

Figure 4. The hysteresis rules of constitutive relation of concrete.

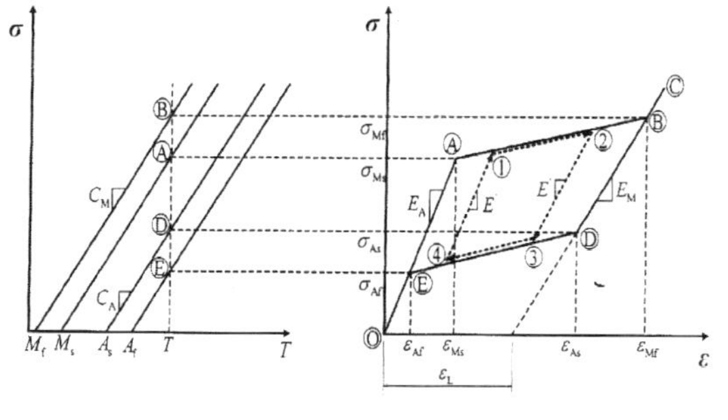

Figure 5. The tensile part of the concrete skeleton curve.

In this model, the degradation and decline of the yield strength of steel skeleton curve considers not only the deterioration of the steel itself, but the effect of reflecting the integrated bond slip degradation of reinforced concrete and concrete interface peeling caused by a protective layer.

3.4. The constitutive model of SMA

The model is based on the SMA thermodynamic model framework derived from Lagoudas [19] et al. It is based on the concept of free energy and dissipation. The model assumes that the loading and unloading phases of each SMA stress and strain are linear, and a multi-linear model is proposed and simplified, greatly reducing the computation time and being more convenient in engineering applications. Figure 5 is the schematic diagram of the superelastic SMA multi-linear constitutive model. C_M and C_A are the critical stress-temperature conversion coefficients of martensitic transformation and reverse phase transformation, respectively. The ε_L is the maximum phase transition strain, and T is the ambient temperature.

3.5. Unit selection and model establishing

The numerical analysis is based on the ABAQUS platform. According to characteristics of RC beam-column joints, a modeling fiber unit is taken [20]. The slip between the reinforcement and the concrete is ignored. In the model analysis, the boundary conditions and loads are imposed by setting two analysis steps. The specific processes are: (1) constraining column bottom three directional displacements, an axial pressure is applied on the column top plate. Then, constraining X and Y directional displacements on the top of the column to prevent the component tilt and beam end force. (2) In the second analysis step, the Z-direction vertical load and the line displacement are applied to both ends of the beam to realize the quasi-static test.

3.6. Loading scheme

In the FEA simulation, the loading system is used, shown in Figure 6. Each cycle includes a forward loading-unloading and a reverse loading-unloading two stages. At start, three cycles are controlled by the vertical load, and the control value is typically 30%, 50% and 70% of the maximum vertical load, respectively. The fourth cycle is changed to the vertical displacement control when the model reaches the yield load. Then the displacement control is used until the model fails.

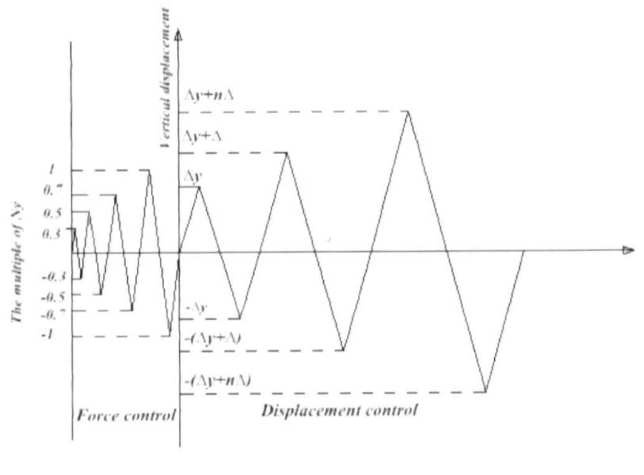

Figure 6. Quasi-static loading scheme.

4. Finite element analysis results

Based on the above modeling method, the simulation results show that the hysteresis performance and the self-centering performance of the pre-stressed beam-column joints of YJD-1, YJD-2, YJD-3, YJD-4, YJD-5 and the ordinary beam-column of XJD-1 are compared.

4.1. Hysteresis curves

After post-processing, beam-force-displacement hysteresis curves of YJD-1and XJD-1 models are obtained by extracting peak displacements in the last hysteresis, shown in Figure 7. The maximum reaction force and residual deformation at the beam end after unloading equivalent are shown in Table 2.

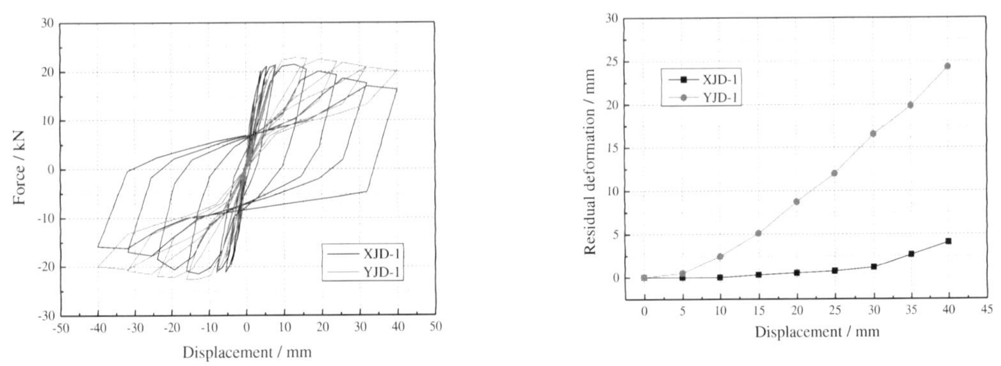

(a) (b)

Figure 7. Comparison of YJD-1 and XJD-1 hysteresis curves and residual deformation. (a) Force-displacement curves. (b) Residual deformation-displacement curves.

Table 2. Comparison of maximum and residual deformation of models.

Specimen number	Maximum deformation (mm)	Maximum reaction force (kN)	Residual deformation (mm)	Residual deformation rate (%)
XJD-1	38.83	21.49	24.28	62.53
YJD-1	39.65	22.83	4.07	10.26
YJD-2	39.61	23.62	3.55	8.96
YJD-3	39.68	24.33	2.95	7.43
YJD-4	39.73	26.90	1.41	3.55
YJD-5	39.71	30.27	0.76	1.91

(1) It can be seen from Figure 7(a) that the vertical reaction-displacement curve is almost linear at the beginning of loading and the hysteresis curve loop area is very small. This is because the initial stiffness of the joint is large and the structure is in the elastic phase. As continuing loading, concrete reaches the ultimate bearing capacity, and the longitudinal reinforcement tends to carry the main load until the steel reaches yield strength. The ultimate load of the model cannot be able to maintain, and the beam bending failure happens in the end.
(2) It can be seen from Figure 7 (a) and (b) that the normal joint and the pre-stressed SMA joints have good energy dissipation capacity. The hysteresis curve of the common beam-column joint shows a long quadrilateral, and the residual deformation is larger after unloading. The hysteresis curve of pre-stressed SMA tendons is round and full, showing a "double flag" curve with typical self-centering characteristics. It can be seen from Table 2 that the residual deformation of the common beam-column joint increases with the loading of the load, and the residual deformation rate is 62.53% in the last loading and unloading stage. In each loading and unloading stage, the residual deformation of the pre-stressed SMA tendons does not change much, and the residual deformation rate is only 10.26% during the last loading and unloading phase, showing good self-centering performance.

4.2. Skeleton curves

In this paper, the envelope curve formed by connecting these points is the skeleton curve of the reciprocating load of the model. The skeleton curves of the XJD-1 and YJD-1 models shown in Figure 8.

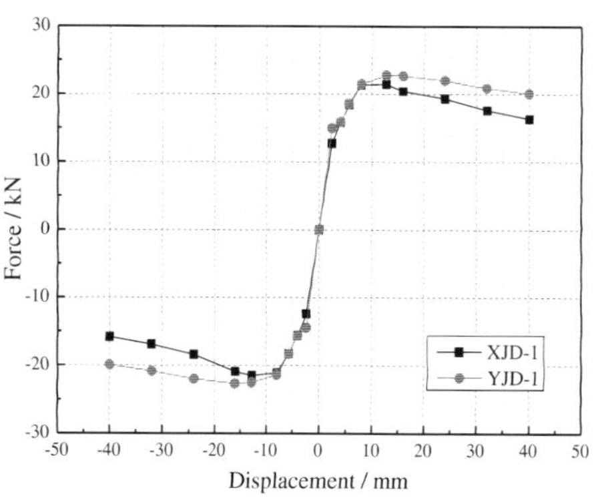

Figure 8. The comparison of skeleton curves between XJD-1 and YJD-1 models.

It can be seen from Figure 8 that the skeleton curves of the two joints generally show "S" shape, which has experienced the elastic stage and the elastoplastic stage. Throughout the process, the beam reaction force of the YJD-1 model is larger than that of the XJD-1 one. With increase of the loading displacement, the two curves separate to form two curves, and both curves have downward trend. When loading to the maximum displacement, two loading capacities are different.

4.3. Performance analysis of hysteretic curve under different influencing factors

4.3.1. The SMA pre-stress level influences. The SMA pre-stress level has an influence on the dynamic performance of the joint. Models of YJD-1, YJD-2 and YJD-3 are respectively subjected to different pre-stress levels by using the cooling method in the FEA model, and other parameters are the same. The hysteresis curves of the beam-side reaction-displacement are obtained, shown in Figure 9.

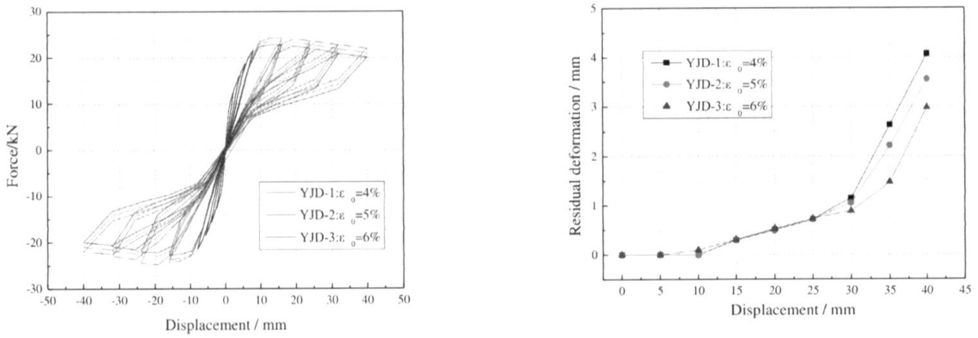

(a) (b)

Figure 9. The comparison of YJD-1, YJD-2 and YJD-3. (a) The hysteresis curves. (b) The residual deformation.

It can be seen from Figure 9 (a) that the magnitude of the initial pre-stress has a significant effect on the joint stiffness but small effect on bearing carrying capacities. Three sets of models are smooth "double flag" shape curves with a high degree of similarity. With increase of loading, SMA tendons are gradually elongated, and the SMA stress on initial stage continues increasing. At each unloading, the SMA tendons pull the beam back to original position due to the superelasticity, and the cracks in the concrete beam become self-closing.

It can be seen from Figure 9 (a), (b) and Table 2 that the initial pre-stress of the SMA tendon is large and the residual deformation of the specimen is small after unloading. As the load continuing increasing, the residual deformation after unloading at each stage gradually decreases as the pre-stress of the SMA tendons increases. It is shown that the pre-stress level tendon has some influences on the reset performance of joints.

4.3.2. The SMA diameter influences. The SMA diameter also has an effect on the dynamic performance of the joint. By changing the diameter of SMA tendons in YJD-3, YJD-4 and YJD-5, retaining the same parameters, the hysteretic displacement hysteresis curves of the beam end are obtained, shown in Figure 10.

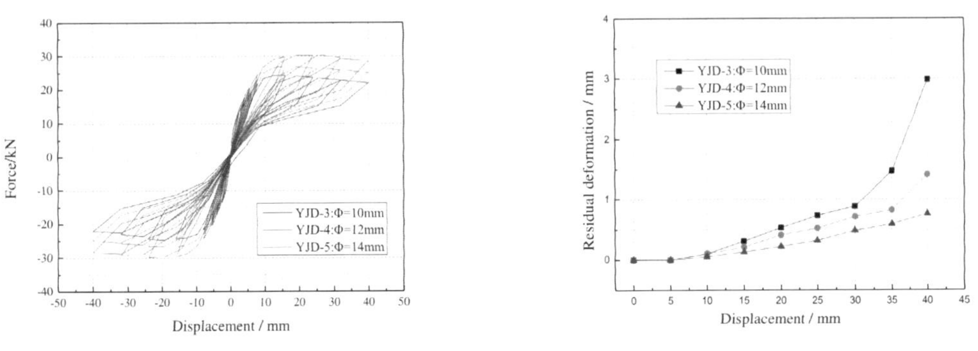

(a) (b)

Figure 10. The comparison of YJD-3, YJD-4 and YJD-5. (a) The hysteresis curves. (b) The residual deformation.

It can be seen from Figure 10(a) that with increase of the SMA diameter, the joint stiffness and bearing capacities of beam-column joints are gradually increased. When the load gradually increases, the SMA tendons are gradually elongated. When the load increases to a certain extent, the main bending stiffness of the model comes from the SMA tendon.
It can be seen from Figure 10(a) and (b) and Table 2 that the residual deformation gradually decreases with increase of the SMA tendon diameter, that is, the self-repairing performance of the joints is getting better.

4.4. Performance analysis of skeleton curves under different influencing factors
The peak points of the beam-side responses are extracted in each cycle of the hysteresis curve in each series, and the envelope curves are formed by connecting these points together, shown in Figure 11. The skeleton curves have the following properties:

- The trend of skeleton curves of joints is basically as the same as the typical "S" shape. The development trend of the model is from steep to slow. The bearing capacities of the models are gradually increased.
- It can be seen from Figure 11(a) that the beam end magnitude in descending order is YJD-3, YJD-2 and YJD-1 in the loading process. It is shown that with the increase of pre-stress level, the reaction force of the beam is gradually increasing.
- It can be seen from Figure 11(b) that the skeleton curve increases with increase of SMA tendon diameter. Therefore, the SMA tendon diameter has a significant effect on the bearing capacity of the joint.

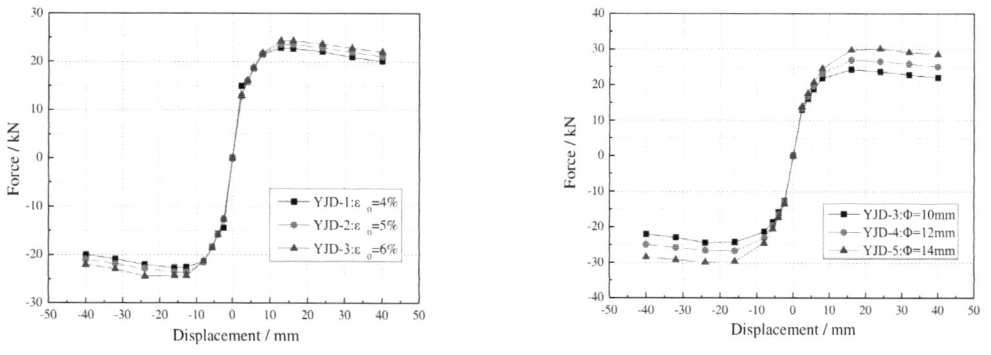

(a) (b)

Figure 11. The skeleton curve comparison for different pre-stress levels and diameters. (a) YJD-1 (4%), YJD-2 (5%) and YJD-3 (6%). (b) YJD-3 (10 mm), YJD-4 (12 mm) and YJD-5 (14 mm).

5. Conclusions
In this paper, by means of finite element method, the hysteretic characteristics, bearing capacity and self-centering ability of the RC beam-column joint reinforced with pre-stressed SMA tendons are analyzed. The main conclusions are as follows:

- The pre-stressed SMA tendons can not only improve the bearing capacity of the joints, but also the self-centering performance of the joints due to the superelastic effect of SMA material.
- The larger the initial pre-stressed of the SMA in the beam-column joints, the better the self-centering ability of the joints, the greater the bearing capacity, and the smaller the residual deformation after unloading.
- In the beam-column joints, with the increase of the SMA diameter, the better the self-centering ability of the joints, the seismic bearing capacity of the joints is obviously improved, and the residual deformation is smaller after unloading.

6. References

[1] Farmani M A and Ghassemieh M 2016 Shape memory alloy-based moment connections with superior self-centering properties. *Smart Materials and Structures* **25** 075028

[2] Shajil N, Srinivasan S M and Santhanam M 2016 An experimental study on self-centering and ductility of pseudo-elastic shape memory alloy (PESMA) fiber reinforced beam and beam-column joint specimens. *Materials and Structures* **49** 783-793

[3] Wang W, Chan T M, Shao H Land Chen Y Y 2015 Cyclic behavior of connections equipped with NiTi shape memory alloy and steel tendons between H-shaped beam to CHS column. *Engineering Structures* **88** 37-50

[4] Song L L, Guo T, Gu Y and Cao Z L 2015 Experimental study of a self-centering prestressed concrete frame subassembly. *Engineering Structures* **88** 176-188

[5] Qian H, Li H N, Song G B 2008 Energy dissipation system of structures with shape memory alloy damper. *Journal of Vibration & Shock* **27** 42-47 (in Chinese)

[6] Moradi S and Alam M S 2015 Feasibility study of utilizing superelastic shape memory alloy plates in steel beam–column connections for improved seismic performance. *Journal of Intelligent Material Systems and Structures* **26** 463-475

[7] Wang W, Yan S, Song G B and Jiao L 2009 An improved two-dimensional constitutive law for shape memory alloys. *Key Egineering Materals* **410-411** 429-437

[8] Yan S, Niu J, Mao P, Song G B and Wang W 2013 Experimental research on passive control of steel frame structure using SMA wires. *Mathematical Problems in Engineering* Article ID: 416282

[9] Ren W, Li H, Song G 2007 Phenomenological modeling of the cyclic behavior of superelastic shape memory alloys. *Smart Materials and Structures* **16** 1083

[10] Huang B, Cai W, Zhao W, et al. 1997 The effect of heat-treatment and cold deformation on non-linear superelasticity of Ti-Ni alloy. *Aerospace Materials & Technology* **6** 51-54 (in Chinese)

[11] Piedboeuf M C, Gauvin R, Thomas M 1998 Damping behaviour of shape memory alloys: strain amplitude, frequency and temperature effects. *Journal of Sound & Vibration* **214** 885-901

[12] Dolce M, Cardone D 2001 Mechanical behavior of shape memory alloys for seismic applications Ⅱ, austenite NiTi wires subjected to tension. *International Journal of Mechanical Sciences* **43** 2657-2677

[13] Ip K H 2000 Energy dissipation in shape memory alloy wires under cyclic bending. *Smart Materials and Structures* **9** 653-659

[14] Sakai Y, Kitagawa Y, Fukuta T, Iiba M 2003 Experimental study on enhancement of self-restoration of concrete beams using SMA wire. *International Society for Optics and Photonics* **5057** 178-186

[15] Chen W 2012 *Applicability of hysteretic constitutive models of concrete in structural analysis based on development of OpenSees*. Chongqing University, Chongqing, China (in Chinese)

[16] Scott B D, Park R, Priestley M J N 1989 *Stress-strain behavior of concrete confined by overlapping hoops at low and high strain ratio Rates*. Doctoral Thesis, Lulea University of Technology, Lulea, Sweden

[17] Yassion M H M 1995 *Nonlinear analysis of prestressed concrete structures under monotonic and cyclic loads*. Ph. D. University of California of Berkeley

[18] Gao X, Zhang Y 2014 Nonlinear analysis of a RC column under cycling loads. *Structural Engineers* **30** 56-63 (in Chinese)

[19] Lagoudas D C, Bo Z, Qidwai M A, et al. 2003 *SMA um: user material subroutine for thermomechanical constitutive model of shape memory alloys*. Texas A & M University, College Station, Texas, USA

[20] Ge D, Wan J, Pan P, et al. 2013 Comparative study on simulation methods of steel bar in fiber beam element model. *Building Structure* **43** 60-62 (in Chinese)

Acknowledgements
This work was partially funded by the Natural Science Foundation of Liaoning Province of China grant 2015020595; the Liaoning Provincial Department of Education Science Research Project grant LJZ2016029; the National Natural Science Foundation of China grant 51278313.

IOP Conf. Series: Materials Science and Engineering **348** (2018) 012014 doi:10.1088/1757-899X/348/1/012014

Physical-durable performance of concrete incorporating high loss on ignition-fly ash

Trong-Phuoc Huynh[1], **Si-Huy Ngo**[2] **and Chao-Lung Hwang**[3]

1 Department of Rural Technology, College of Rural Development, Can Tho University, Campus II, 3/2 St., Ninh Kieu Dist., Can Tho City 900000, Vietnam
2 Department of Engineering and Technology, Hong Duc University, No. 565, Quang Trung Street, Dong Ve Ward, Thanh Hoa City 440000, Viet Nam
3 Department of Civil and Construction Engineering, National Taiwan University of Science and Technology, No. 43, Sec. 4, Keelung Rd., Taipei 10607, Taiwan
E-mail: htphuoc@ctu.edu.vn

Abstract. This study investigates the feasibility of using raw fly ash with a high loss on ignition in concrete. The fly ash-free concrete samples were prepared with different water-to-binder (w/b) ratios of 0.35, 0.40, and 0.45, whereas the fly ash concrete samples were prepared with a constant w/b of 0.40 and with various fly ash contents (10%, 20%, and 30%) as a cement substitution. The physical properties and durability performance of the concretes were evaluated through fresh concrete properties, compressive strength, strength efficiency of cement, ultrasonic pulse velocity, and resistance to sulfate attack. Test results show that the w/b ratio affected the concrete properties significantly. The incorporation of fly ash increased the workability and reduced the unit weight of fresh concrete. In addition, the fly ash concrete samples containing up to 20% fly ash exhibited an improved strength at long-term ages. Further, all of the fly ash concrete samples showed a good durability performance with ultrasonic pulse velocity value of greater than 4100 m/s and a comparable sulfate resistance to the no-fly ash concrete.

1. Introduction

With the rapid development of construction industry, ordinary Portland cement concrete has become the most popular construction material in the world. Therefore, a large amount of cement is produced and consumed every year and thus the generation of carbon dioxide (CO_2) during the production of ordinary Portland cement is consequently a serious environmental problem, causing the greenhouse effect and global warming. The production of ordinary Portland cement also consumes a significant energy and depletes the natural resources. Thus, it is necessary to reduce the emission of CO_2 and the disposal of industrial wastes, in which fly ash, a by-product from coal power plant, has been utilizing as a partial or full replacement of ordinary Portland cement in concrete instead of considering it as an industrial waste. Although fly ash has been widely used in concrete for a long time, however, the results from previous studies varied depending on the source and characteristic of fly ash used. Wankhede and Fulari [1] stated that concrete with 10% and 20% fly ash replacing cement showed a better compressive strength than fly ash-free concrete, but the inverse result was observed with 30% fly ash content. Bansal et al. [2] reported that compressive strength of concrete containing 20% and 30% fly ash was about 11% and 19% higher than that of the no-fly ash concrete, respectively. Marthong and Agrawal [3] investigated the utilization of 10%, 20%, 30%, and 40% fly ash replacing cement in concrete. Their results exhibited that increasing the fly ash content yielded the reduction in compressive strength of concrete. However, the concrete with 20% fly ash content had a comparable compressive strength to the conventional concrete without fly ash. Naik and Ramme [4] and Oner et al. [5] studied the use of up to 60% fly ash in concrete as a cement replacement. The results of these studies showed that the optimum fly ash replacement level was about 40%, at which showing the improved compressive strength. However, Siddique [6] indicated that concrete containing 40–60% fly ash had lower strength than normal concrete without fly ash. On the other hand, Li and Zhao [7] found that the combination of 25% fly ash and 15% granulated blast furnace slag enhanced both short- and long-term properties of concrete. Previous studies also reported that the workability of concrete increased with increasing the fly ash replacement level [1,3,4] and the use of fly ash in concrete also improved the resistance to sulfate attack [7,8].

Generally, the use of fly ash in concrete was found to have a positive effect on the concrete properties. However, the optimum dosages of fly ash used in concrete varied depending on the sources of fly ash and its quality. The selected fly ash with the loss on ignition of less than 3.2% was used in most of the previous studies [4–8]. Thus, this study aims to investigate the feasibility of using raw fly ash with a high loss on ignition in concrete in order to fill the gap in the literature. The fly ash used in this study had the loss on ignition of 15.7%, which is much higher than the suggested value of lower than 6% of the ASTM C618 [9]. The effect of such fly ash on properties of both fresh and hardened concrete was investigated in the present study.

2. Materials and experimental works

2.1. Materials properties and mixture proportions

All of the materials used in this study were locally sourced from Thanh Hoa province, Vietnam. Nghi Son type-PC40 cement and class-F fly ash with the respective specific gravity of 3.12 and 2.16 were used as binder materials. Table 1 shows the chemical compositions of both cement and fly ash. It is noted that the fly ash used in this study is a raw material, which was taken from a local thermal power plant, with a high loss on ignition value of 15.7%. Natural sand and crushed stone, with properties as shown in Table 2, were used as fine and coarse aggregates, respectively. The superplasticizer (SP) with a specific gravity of 1.15 was used to obtain the desired workability of fresh concrete with less water.

The ingredient proportions of all concrete mixtures are shown in Table 3. First three concrete mixtures (Group I) were designed with different w/b ratios of 0.35, 0.40, and 0.45. The group I samples used to examine the effect of the w/b ratio on physical and durable properties of the concrete. The last three mixtures (Group II) were designed with a constant w/b ratio of 0.40 and various fly ash replacement levels of 10%, 20%, and 30% cement. The group II samples used to investigate the effect of fly ash content on physical and durable properties of the concrete. In this study, all of the concrete mixtures were designed in accordance with ACI 211.91 [10] with controlling the amount of water less than 180 kg/m^3 for good durability as suggested by Hwang and Hung [11]. Nomenclature of the mixtures is described as follows: M35, M40, and M45 denote the respective w/b ratios of 0.35, 0.40, and 0.45. The following numbers (0, 10, 20 and 30) are the percentages of fly ash replacing cement in concrete mixtures.

Table 1. Chemical compositions of cement and fly ash.

Compositions (wt. %)	SiO_2	Al_2O_3	Fe_2O_3	CaO	MgO	Others	Loss on ignition
Cement	22.4	5.3	4.0	55.9	2.8	4.4	2.0
Fly ash	48.4	20.4	4.8	2.8	1.4	4.3	15.7

Table 2. Properties of aggregates used in this study.

Aggregate types	Maximum size (mm)	Fineness modulus	Density (T/m^3)	Dry rodded weight (T/m^3)	Moisture content (%)	Water absorption (%)
Natural sand	5.0	2.87	2.62	1.50	4.35	1.08
Crushed stone	12.5	-	2.69	1.39	0.25	0.08

Table 3. Mix proportions for the preparation of concrete samples.

Sample ID.	w/b	Concrete ingredients proportion (kg/m^3)					
		Cement	Fly ash	Sand	Stone	Water	SP
M35-0	0.35	514.1	0.0	985.2	751.1	175.5	5.1
M40-0	0.40	450.0	0.0	1039.0	751.0	176.1	4.5
M45-0	0.45	400.4	0.0	1082.1	751.8	175.7	4.0
M40-10		402.4	44.7	1032.4	746.2	175.0	4.5
M40-20	0.40	355.4	88.9	1025.9	741.5	173.9	4.4
M40-30		309.1	132.5	1019.4	736.8	172.8	4.4

2.2. Samples preparation and test methods
Right after mixing, the properties of fresh concrete such as slump, unit weight were checked. The cylindrical concrete samples with a diameter of 100 mm and a height of 200 mm were prepared for testing of compressive strength, ultrasonic pulse velocity, and resistance to sulfate attack. In order to evaluate the resistance to sulfate attack, concrete specimens were fully immersed in 5% Na_2SO_4 solution right after demolding until the testing days. All of these tests were conducted at 3, 7, 14, 28, 56, and 91 days, with the reported results of each test are the average value of three concrete samples. Slump, compressive strength, and ultrasonic pulse velocity tests were performed in accordance with ASTM C143 [12], ASTM C39 [13], and ASTM C597 [14], respectively.

3. Results and discussion

3.1. Properties of fresh concrete
Fresh concrete properties of all mixtures are given in Table 4. As shown, the slump and fresh unit weight of fresh concrete mixtures ranged from 0–50 mm and 2470–2580 kg/m^3, respectively. For the no-fly ash group, M45-0 mixture showed the highest slump, followed by M40-0 and M35-0. An inverse trend was observed in term of unit weight, in which M35-0 mixture exhibited the highest fresh unit weight, followed by M40-0 and M45-0. In this group, the slump of fresh concrete is closely related to the amount of water. So that when the w/b ratio increased, the amount of water increased. Thus, the workability of fresh concrete consequently improved. On the other hand, a low w/b ratio is associated with a higher amount of cement used in the concrete mixture. Hence, the higher unit weight of fresh concrete is mainly attributable to the higher specific gravity of cement compared with that of other ingredients in concrete mixtures.

Table 4. Properties of fresh concrete mixtures.

Sample ID.	w/b	SP dosage (kg/m^3)	FA content (%)	Slump (mm)	Fresh unit weight (kg/m^3)
M35-0	0.35	5.1	0	0	2580
M40-0	0.40	4.5	0	20	2550
M45-0	0.45	4.0	0	30	2530
M40-10		4.5	10	20	2530
M40-20	0.40	4.4	20	35	2500
M40-30		4.4	30	50	2470

Table 4 also shows that increasing the fly ash replacement level resulted in an increase in workability and a reduction in unit weight of fresh concrete mixtures. The increased workability is mainly due to the spherical shape of fly ash particles. It is well known that the particle shape of cement and fly ash are irregular polygonal and spherical, respectively [15]. Therefore, owing to a spherical shape, the presence of fly ash reduced the friction between the aggregate and paste, thus improving the workability of fresh concrete. Moreover, due to the lower specific gravity of fly ash as compared with that of cement, the volume of fly ash paste is greater than that of the cement paste. The improvement of the plasticity and cohesion of the paste led to increasing the workability of fresh concrete. This finding is supported by the results of the previous studies [1,3,4]. On the other hand, the unit weight of fresh concrete reduced with increasing fly ash content in the concrete mixture. This is also due to the lower specific gravity of fly ash as compared with that of cement.

3.2. Compressive strength development

The effects of w/b ratio and fly ash content on compressive strength development of concrete are shown in Figure 1 and Figure 2, respectively. The 91-day compressive strength values of the M35-0, M40-0, and M45-0 mixtures were 39.5, 30.0, and 26.3 MPa, respectively. As the results, the concrete samples with lower w/b ratio had higher compressive strength values. With a similar amount of water, a concrete mixture with low w/b ratio used more amount of cement, resulting in a higher cement hydration rate and thus more hydration product as expressed in Equation (1), which contributed to a higher strength.

On the other hand, concrete samples incorporating 10% fly ash showed the highest compressive strength at the early ages (before 28 days), followed by the concrete samples with 0%, 20%, and 30% fly ash. However, after 56-day ages, concrete samples containing 10% and 20% fly ash exhibited the higher compressive strength than fly ash-free concrete. As mentioned in previous studies [1-7], the compressive strength of fly ash concrete was enhanced with the use of an optimum amount of fly ash. In case of using fly ash over the optimum level, a part of fly ash does not take part in the chemical reaction process. It mainly acts as a filler or fine aggregate rather than a supplementary cementitious material. The optimum amount of fly ash used in this study was around 20%, which is lower than that presented in previous studies [4–7] due to the poor quality of the raw fly ash used as aforementioned. The lower compressive strength at short-term and higher compressive strength at long-term are associated with the pozzolanic reaction of fly ash, which only significantly started after several weeks of the sample after casting. The continuous increase of compressive strength of fly ash concrete is due to the transformation of $Ca(OH)_2$ into the secondary C-S-H gels followed the Equation (2), which are the main carriers of loading in hardened concrete. Although M40-30 mixture with the use of 30% fly ash content showed the lowest compressive strength, its compressive strength was approximate 88% that of control mixture. All of the fly ash concrete mixtures showed a compressive strength value at 91 days of above 26.5 MPa. This finding further proved the feasibility of using the raw fly ash with a high loss on ignition in concrete.

Cement hydration:

$$(C_3S, C_2S) + H_2O \rightarrow C\text{-}S\text{-}H + Ca(OH)_2 \qquad (1)$$

Pozzolanic reaction:

$$Ca(OH)_2 + SiO_2 \rightarrow C\text{-}S\text{-}H \qquad (2)$$

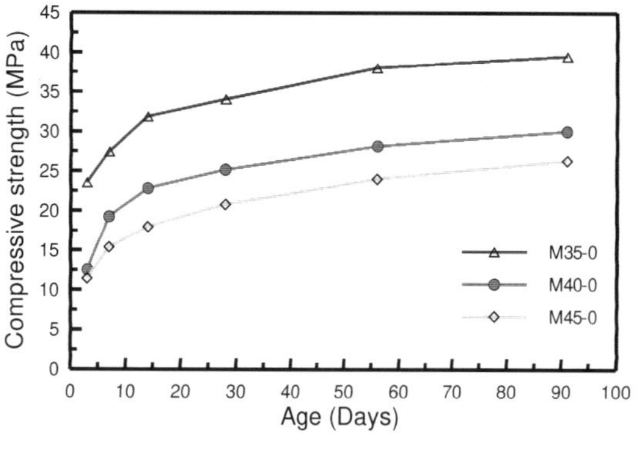

Figure 1. Effect of w/b on compressive strength development of the concrete samples.

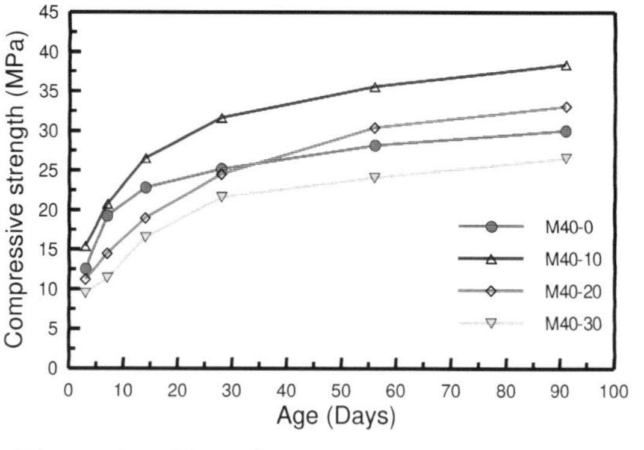

Figure 2. Effect of FA content on compressive strength development of the concrete samples.

3.3. Strength efficiency of cement

The strength efficiency of cement (SEC) is a yielded concrete compressive strength per kilogram of cement. So far, the amount of cement used in each mixture is associated with the production cost, energy consumption, and CO_2 emission during the production of cement and concrete. Thus, the SEC is an indicator for cost-effectiveness and environmental reduction of fly ash concrete mixtures. Figures 3 and 4 show the SEC in concrete samples prepared with various w/b ratios and fly ash contents, respectively. As can be seen from Figure 3, M35-0 mixture had the highest SEC value, followed by M40-0 and M45-0. This is due to the high volume of cement paste in the concrete mixture at the low w/b ratio, resulting in a greater hydration rate and thus more cement hydration products.

As shown in Figure 4, the 91-day SEC value of the M40 concrete mixtures with 0%, 10%, 20%, and 30% fly ash content were 0.067, 0.095, 0.092, and 0.084 MPa/kg/m^3, respectively. The use of fly ash in concrete increased the SEC of about 26–42% as compared with the control concrete mixture (M40-0). This indicated that most fly ash particles participated in the pozzolanic reaction rather than existed as a filler or a fine aggregate. Fly ash particles reacted with calcium hydroxide to form C-S-H gel as expressed in Equation (2), thus improving the compressive strength of concrete. This also proved the cost-effectiveness when using this low-quality fly ash in concrete.

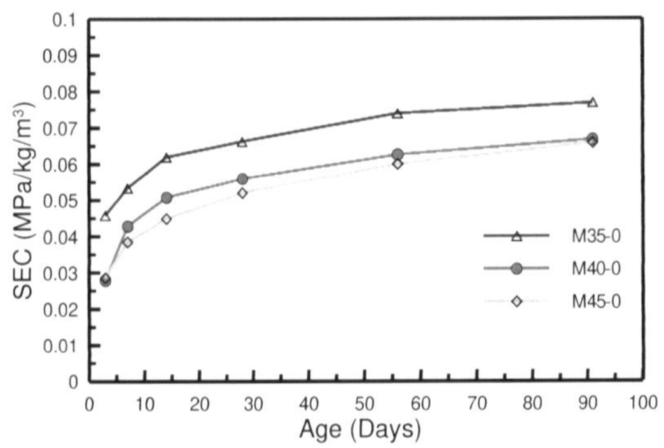

Figure 3. Effect of w/b on strength efficiency of cement in concrete samples.

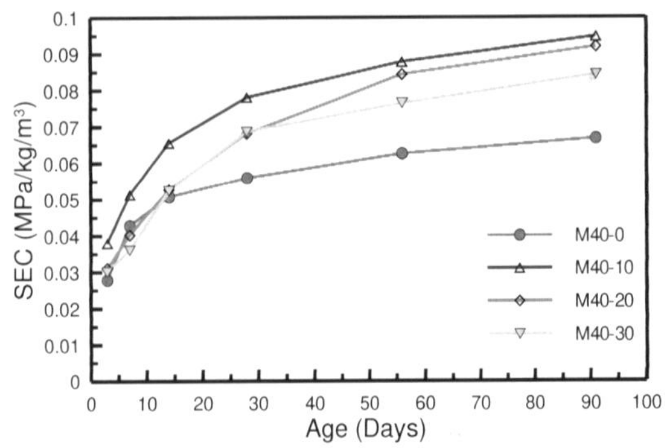

Figure 4. Effect of fly ash content on strength efficiency of cement in concrete samples.

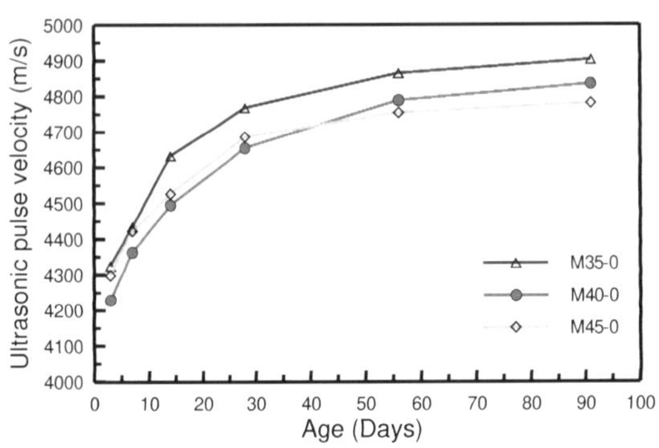

Figure 5. Effect of w/b on UPV values of the hardened concretes.

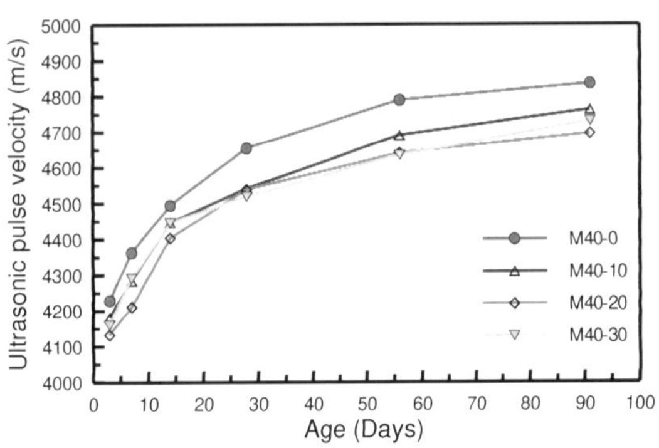

Figure 6. Effect of fly ash content on UPV values of the hardened concretes.

3.4. Ultrasonic pulse velocity

Values of the ultrasonic pulse velocity (UPV) is one of the indicators for durability of hardened concrete. The durable properties of concrete are often related to the structure of hardened concrete as the uniformity and the presence of voids and cracks. In general, concrete with a high value of UPV is associated with a denser structure and a good quality. Figures 5 and 6 show the results of the UPV test of concrete samples prepared with different w/b ratios and various fly ash replacement levels, respectively. As the results, the

UPV of all of the concrete samples increased continuously with the curing ages. Malhotra [16] previously reported that concrete with UPV value of higher than 3660 m/s was in good condition and durable. Thus, according to Malhotra suggestion, all of the concrete samples prepared in this study exhibited a good quality and durability with the UPV values of above 4100 m/s.
Figure 5 clearly shows that the 91-day UPV value of M35-0 mixture was the highest, followed by M40-0 and M45-0, whereas Figure 6 shows that increasing the level of fly ash replacement resulted in a decrease in UPV values. This is attributable to the amount of cement used in the mixtures as well as the density of concrete structure. The development of UPV value is due to the combined cement hydration and pozzolanic reaction took place within the system during the time.

3.5. Sulfate attack

The durability properties of concrete are greatly affected by the environmental condition and especially by the chemical attack. One of the most aggressive environmental agents is sulfate, which is popularly existed in soil, groundwater, and seawater in some areas. The compressive strength test results of concrete samples that fully immersed in 5% Na_2SO_4 solution are given in Table 5. As the results, the compressive strength values of all of the concrete samples in sulfate solution still kept increasing with the immersion time. Similar to the concrete samples cured in water, concrete in Na_2SO_4 solution with a lower w/b ratio showed a higher compressive strength and concrete with 10% and 20% fly ash content exhibited the higher compressive strength than that of the concrete without fly ash.

Table 5. The compressive strength of concretes immersed in 5% Na_2SO_4 solution.

Sample ID.	Compressive strength (MPa)					
	3 days	7 days	14 days	28 days	56 days	91 days
M35-0	20.64	26.07	28.41	29.40	31.17	32.51
M40-0	13.35	19.03	22.24	24.80	25.89	27.02
M45-0	11.73	13.90	15.80	20.24	23.07	23.91
M40-10	15.93	17.94	24.53	29.68	32.38	35.15
M40-20	11.19	13.19	18.36	22.18	26.45	29.87
M40-30	9.58	12.43	16.15	18.91	21.27	23.63

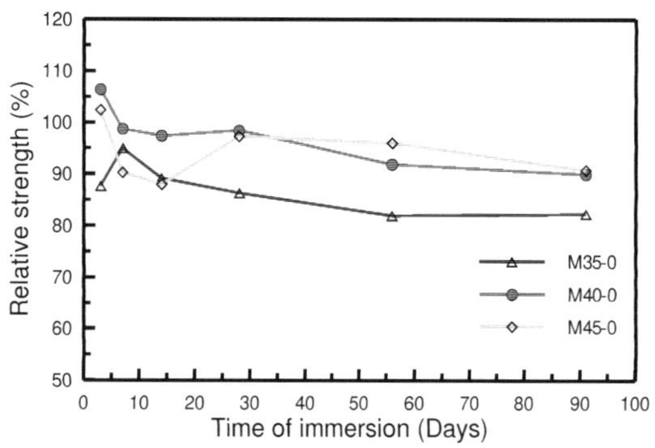

Figure 7. Effect of w/b on sulfate resistance of the hardened concretes.

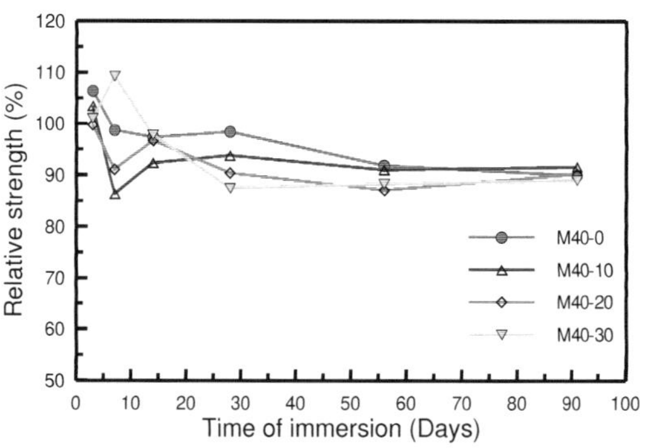

Figure 8. Effect of fly ash content on sulfate resistance of the hardened concretes.

In order to evaluate and compare the development of compressive strength between concrete samples cured in water and the samples immersed in sulfate solution, the relative compressive strength values between these samples are plotted as shown in Figures 7 and 8. At the early ages, some samples obtained the relative strength of higher than 100%. This is due to the continuous cement hydration as expressed by Equation (1) or the reaction of fly ash with the liberated lime to form the C-S-H as expressed by Equation (2), and the reaction of sulfate ions with hydration products to form gypsum and ettringite as expressed by Equations

(3)–(6). The formation of C-S-H, gypsum, and ettringite lead to a denser structure and thus increasing the compressive strength of concrete under sulfate solution. However, after several weeks, the sulfate attack reaction become more dominant, the C-S-H was transformed to N-S-H and the formation of micro-cracks due to the expansion of gypsum and ettringite led to a reduction in concrete strength.

$$Na_2SO_4.2H_2O + Ca(OH)_2 \rightarrow CaSO_4.2H_2O + 2NaOH \quad (3)$$

$$Na_2SO_4.2H_2O + \text{C-S-H} \rightarrow CaSO_4.2H_2O + \text{N-S-H} \quad (4)$$

$$C_4AH_{13} + 3CaSO_4.2H_2O + 14H_2O \rightarrow C_3A.3CaSO_4.32H_2O + Ca(OH)_2 \quad (5)$$

$$C_3A.CaSO_4.12H_2O + 2CaSO_4.2H_2O + 16H_2O \rightarrow C_3A.3CaSO_4.32H_2O \quad (6)$$

As can be seen from Figure 7 that the reduction in strength of M35-0 mixture was the most significant with an approximation of 18% at 91-day of immersion. This is attributable to the high amount of cement, resulted in high potential shrinkage and micro-cracks. Whereas, the M40-0 and M45-0 mixtures showed a similar rate of reduction in concrete strength (around 10%) at 91-day of immersion. On the other hand, all of the concrete samples shown in Figure 8 presented the similar strength reduction rate at the later ages of the samples. As previously reported by Li and Zhao [7] and Torii et al. [8], the use of fly ash improved the resistance of concrete to sulfate attack. It is noted that those studies used selected fly ash with the loss on ignition less than 3.2%. However, the poor quality raw fly ash with a high loss on ignition of 15.7% was used in the present study. The fly ash concrete prepared for this study still exhibited a comparable ability of sulfate resistance compared with no fly ash concrete. This finding demonstrated the feasibility of using fly ash with a high loss on ignition in concrete under sulfate attack condition.

4. Conclusions

This research work investigates the physical and durable performance of concrete incorporating high loss on ignition-fly ash. Based on the experimental results, some main conclusions may be drawn as follows:
1. The workability of fresh concrete increased while the fresh unit weight reduced with increasing the fly ash replacement level.
2. The compressive strength of concrete increased with up to 20% fly ash content in the mixture, with a maximum compressive strength value of 41.3 MPa. In addition, the fly ash concrete had 26–42% higher strength efficiency of cement than that of the fly ash-free concrete.
3. All of the fly ash concrete mixtures showed the good durability performance with the UPV values of above 4100 m/s and a comparable sulfate resistance to the no-fly ash concrete.
4. The physical-durable performance of concrete was also significantly affected by the w/b ratio as higher w/b ratio resulted in higher workability, however, lower values of fresh unit weight, SEC, UPV, and compressive strength of the concrete.
5. The results of the present study further demonstrate a high possibility of using of fly ash with a high loss on ignition in the production of concrete.

5. References

[1] Wankhede P R and Fulari V A Effect of fly ash on properties of concrete 2014 *Int. J. Emerg. Technol. Adv. Eng.* **4** 284–9
[2] Bansal R, Singh V and Pareek R K Effect on compressive strength with partial replacement of fly ash 2015 *Int. J. Emerg. Technol.* **6** 1–6
[3] Marthong C and Agrawal T P Effect of fly ash additive on concrete properties 2012 *Int. J. Eng. Res. Appl.* **4** 1986–91
[4] Nail T R and Ramme B W Effect of high-lime fly ash content on water demand, time of set, and compressive strength of concrete 1990 *ACI Mater. J.* **87** 619–26
[5] Oner A, Akyuz S and Yildiz An experimental study on strength development of concrete containing fly ash and optimum usage of fly ash in concrete 2004 *Cem. Concr. Res.* **35** 1165–71

[6] Siddique R Performance characteristics of high-volume class F fly ash concrete 2003 *Cem. Concr. Res.* **34** 487–93
[7] Li G and Zhao X Properties of concrete incorporating fly ash and ground granulated blast-furnace slag 2003 *Cem. Concr. Compos.* **25** 293–9
[8] Torii K, Taniguchi K and Kawamura M Sulfate resistance of high fly ash content concrete 1995 *Cem. Concr. Res.* **25** 759–68
[9] ASTM C618 2005 *Standard specification for coal fly ash and raw or calcined natural pozzolan for use in concrete*
[10] ACI Committee 211 1991 *Standard practice for selecting proportions for normal, heavyweight, and mass concrete*
[11] Hwang C L and Hung M F Durability design and performance of self-consolidating lightweight concrete *Constr. Build. Mater.* **19** 619–26
[12] ASTM C143 2015 *Standard test method for slump of hydraulic-cement concrete.*
[13] ASTM C39 2012 *Standard test method for compressive strength of cylindrical concrete specimens*
[14] ASTM C597 2009 *Standard test method for pulse velocity through concrete*
[15] Papadakis V G Effect of fly ash on Portland cement systems – Part I. Low-calcium fly ash 1999 *Cem. Concr. Res.* **11** 1727–36
[16] Malhotra V M 1976 *Testing hardened concrete: Nondestructure methods* American Concrete Institute Monograph Series

Variability of chemical analysis of reinforcing bar produced in Saudi Arabia

A. Salman[1], F. Djavanroodi[2]

1Department of Civil Engineering, Prince Mohammad Bin Fahd University, Al Khobar, KSA
2Department of Mechanical Engineering, Prince Mohammad Bin Fahd University, Al Khobar, KSA
E-mail: fdjavanroodi@pmu.edu.sa

Abstract: In view of the importance and demanding roles of steel rebar's in the reinforced concrete structures, accurate information on the properties of the steels is important at the design stage. In the steelmaking process, production variations in chemical composition are unavoidable. The aim of this work is to study the variability of the chemical composition of reinforcing steel produced throughout the Saudi Arabia and asses the quality of steel rebar's acoording to ASTM A615. 68 samples of ASTM A615 Grade 60 from different manufacturers were collected and tested using the Spectrometer test to obtain Chemical Compositions. EasyFit (5.6) software is utilized to conducted statistical analysis. Chemical compositions distributions and, control charts are generated for the compositions. Results showed that some compositions are above the upper line of the control chart. Finally, the analyses show that less than 3% of the steel failed to meet minimum ASTM standards for chemical composition.

1. Introduction

Conventional reinforced concrete is a composite material of reinforcing steel bars embedded in a hardened concrete matrix. The accurate information on the properties of the reinforcing steels as a construction material is important at the design or construction stage. There are many codes which specify the limits on the properties and testing procedures of the steel rebar. These are ASTM A615, BS4449, ISO 6935-2 etc. Saudi Arabia have adopted ASTM A615 standard [1] for the steel rebar testing. Reinforcing bars tests in most construction sites have been restricted to tensile and bend tests with little or no information about chemical composition as they relate to the structural performances of the bars.

Chemical composition variations in producing reinforcing steel bars are unavoidable. Table 1 gives the list of chemical ingredients that influences the property of steel rebar's [2]. Carbon is the main strengthening element that participates in two strengthening mechanisms, solid solution and second phase formation (cementite). Although carbon increasing the strength (mainly the tensile TS), but on the other hand decrease the ductility and Hardness, in addition to affecting the weldability [3]. The carbon equivalent represents the contribution of carbon and other elements to the formation of structures susceptible of hydrogen embrittlment during welding. In the carbon equivalent diagram (Figure 1) the transformation is schematically represented. The rebar steels moved from a region of optimum-medium weldability to region of high risk weldability. The ASTM A615 reflects this problem and explicitly excludes the weldbality. Other properties of the rebar are compromised when the carbon steel with medium carb on content is used for fabricating rebar. Among these is the elastic -

plastic behavior. Most theories of concrete reinforced structure design considers that the steel in the structure deforms elastically and then it will yield plastically at constant stress. The elastic –perfectly plastic behavior of the steel is assumed by most of the model or code for earthquake resistance construction. Metallurgically, hot rolled plain low carbon steel (carbon content less than 0.3 wt. %) produces steel with near elastic-perfectly plastic behavior (i.e. stress – strain curve with wide Luders deformation after yielding). One method to fabricate steel rebar with high strength combined with high ductility, weldability and toughness is Tempcore process. In this method hot rolled deformed bars are quenched at the end of rolling mill by applying high pressure jets of cold water on the red hot steel surface. This process hardens a crust near the steel surface while the bar core remains with high ductility. This procedure produces steel with low carbon, high strength, high ductility, good weldable property and tough material [4].

Table 1. Influence of different chemical ingredients in steel on properties of rebars [2]

Chemicals	**Property**	**Effects on the Rebar's**
Carbon (C)	Hardness, strength, weldability and brittleness	Higher carbon contributes to the tensile strength of steel, that is, higher load bearing capacity and vice versa. Lower carbon content less than 0.1 percent will reduce the strength. Higher carbon content of 0.3 percent and above makes the steel bar unweldable and brittle.
Manganese (Mn)	Strength and yield strength	Higher manganese content in steel increases the tensile strength and also the carbon equivalent property.
Sulphue (S)	It is impurity in steel which increases its brittleness.	Presence of sulphur should be limited. Presence of higher sulphur makes the bar brittle during twisting, as higher sulphur content brings the hot shot problem during rolling
Phosphorus (P)	It is an impurity which increases strength brittleness	Higher phosphorus content contributes to the increase in strength and corrosion resistance properties but brings brittleness due to the formation of low euctoid phosphicles in the grain boundary. Also lowers the impact and value at subzero temperature level (transition temperature).
Copper (Cu)	Strength and corrosion resistance	Being a pearlite stabiliser, it increases the strength and resistance corrosion property
Chromium (Cr)	Weldability and corrosion resistance	Present as an impurity from the scrap and influences carbon equivalent; weldability and increases corrosion resistance property.
Carbon Equivalent (CE or Ceq)	Hardness, tensile strength and weldability	This property is required to set the cooling parameters in TMT (Thermo mechanically treated) process and a slight variation in carbon equivalent may alter the physical properties. In case of CTD (Cold twisted deformed) bars, carbon equivalent has a maximum limit of 0.42 percent but there is no lower limit prescribed. As such, as long as the chemical composition and physical properties of raw materials are within specified limits, the variation in carbon equivalent as in the case of TMT bars.

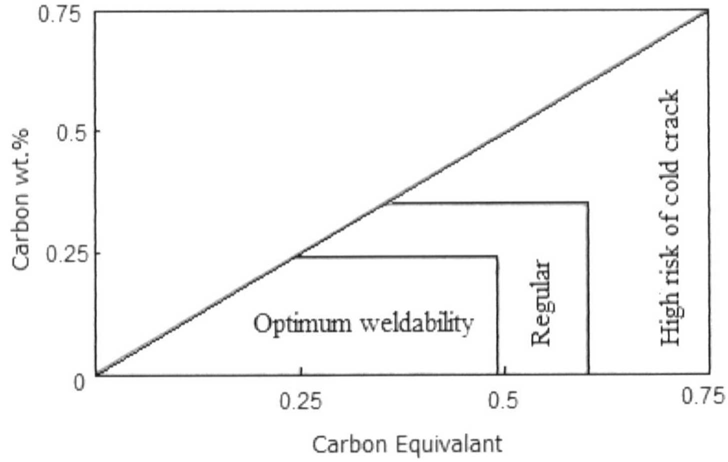

Figure 1. Carbon – carbon equivalent diagram indicating the area of optimum weldability, regular and high risk of cold crack formation [4].

There has been number of statistical studies dealing specifically with the variability of the mechanical properties of reinforcing steel [5-17]. In these studies, variations in yield and tensile strengths were examined. These variations were believed to be caused by variation in the rolling practices and quality control measures used by different manufacturers, as well as possible variations in cross-sectional area, steel strength, and rate of loading. On the other hand there has been very few studies on variability of chemical composition reinforcing steel [18, 19]. Jibrin and Ejeh [19] studied the Chemical composition of reinforcing steel in the Nigerian Construction Industry. A total of 14 companies supplied nineteen samples were tests. Most of the bars showed absence of some element such as Molybdenum, Vanadium, etc which is a strength and coefficient of weldability determinants BS4449. Also, it is shown that the high percentage of elements such as Silicon and Phosphorus impacted negatively on the strength and deformation characteristics of the bars.

Saudi Arabia's steel demand has made the country one of the largest consumer in the GCC region. The country also accounts for significant number of construction activities in the Middle-East region [20]. The purpose of this study is to assess the quality of rebar's manufactured in KSA. The tasks to deliver these objectives will cover the chemical composition testing of the rebars produced from the steel manufacturing plants in the KSA and to derive and investigate the relevant parameters related to the quality of reinforcing bars. This paper reports on the variability of the chemical composition of reinforcing steel produced throughout the Saudi Arabia by experimentation and asses the quality of steel rebar's which satisfy the minimum requirements established by ASTM International A615. Samples ASTM A615 Grade 60 from different manufacturers were collected and tested using the Spectrometer test to obtain Chemical Compositions. A statistical analysis of bar Chemical Composition is conducted.

2. Experimental method

Samples of steel bars ASTM A615 Grade 60 were collected and labeled from eight local steel manufacturing companies in Saudi Arabia as shown in Table 2. Sixtey eight specimens were randomly selected from the manufactures stockpiled. All the samples were prepared and tested for the chemical properties of the steel using a Spectrolab *Foundry Master X-lin* analytical instrument. These tests were performed at three different laboratories namely Saudi Arabia standard organization, SABIC and Imperial College London.

Table 2: List of steel rebars manufactures

Usaimi	SABIC	Al Rajhi	Al Ittefaq	Jazira	Watania	Taybah	Al Yamamah
US	HA	RA	IT	JA	WA	TA	YA

3. Results and discusion

The chemical analysis of the tests carried on the steel samples shows that the steel bars from these manufacturers possessed percentage average Carbon, Phosphors, Sulphuer, Copper, Chromium, Manganese and Carbon Equivalent contents of 0.26, 0.02, 0.013, 0.079, 0.061, 0.852 and 0.402, respectively. EasyFit 5.6 is utilized to obtain necessary statistical results. Table 3 and includes a summury of statistical analysis of chemical compositions. Figure 2 shows full statisctal results of the ten chemical compositions. These compositions follow different types of continuous distributions as depicted in Figure 2. Mean, standard deviation, and other statistical functions are determined according to distribution type. Carbon (C) Copper (Cu), and Manganese (Mn) follow Generalized Extreme Value distribution; Phosphorus (P) follows Log-Logistic distribution; Sulphur (S) follows Log-Pearson distribution; Chrome (Cr) follows Weibull distribution, Molybdenum (Mo) follows Generalized Pareto distribution; Nickel (Ni) follows Fatique Life distribution; Vanadium (V) follows Johnson SB distribution; Silicon (Si) follows Gamma distribution; and Carbon Equivalent (C.E.) follows Dagum. Carbon (C) is the only composition that has left skewed, the other compositions have right sekewed. Phosphorus (P) has highet value of coefficient of variation (CV), which is a function of mean and standard deviation. While, Carbon (C) has the lowest value of CV.

Furthermore, control charts are generated for each chemical composition to verify samples that are below 3 sigma or more than 3 sigma, as shown in Figure 3. Carbon (C), Sulphur (S), Molybdenum (Mo), Vanadium (V), and Manganese (Mn) , and Carbon Equivalent (C.E.) are 100% validation. All values between the upper control line (UCL) and the lower control line (LCL). One sample has Copper (Cu) upper the control line, the validation of Phosphorus (P) is 98%. Finally, two samples have Phosphorus (P), Chrome (Cr), Nickel (Ni), and Silicon (Si) more than the upper control line, the validation of the previous four chemical compositions are 97%. These results are considered acceptable according to ASTM standards.

The result of Spectormeter tests with statistical analysis on steel rebar chemical composition are shown in Table 3. The ASTM A615 does not specify any restriction on Carbon content; the only restriction is on the phosphorus content with a maximum of 0.06 %.and 0.075% from the heat of steel and the product, respectively. The analyses shows that out of the 68 test performed only two samples did not meet the standard i.e. less than 3% of the steel failed to meet minimum ASTM standards for chemical composition. The statistical analysis including the values of maximum, minimum, average, median and standard deviation for each element is calculated and shown in Table 3. The carbon equivalan was calculated using the following equation [21].

$$C.E. = \%\,C + \frac{\%\,Mn}{6} + \frac{\%\,Cu}{40} + \frac{\%\,Ni}{20} + \frac{\%\,Cr}{10} - \frac{\%\,Mo}{50} - \frac{\%\,V}{10}$$

Table 3. Statistical analysis of Chemical Composition of A 615 Grade 60 steel rebar's

Elements	C%	P%	S%	Cu%	Cr%	Mo%	Ni%	V%	Si%	Mn%	C.E.
Distribution Type	Generalized Extreme Value	Log-Logistic	Log-Pearson	Generalized Extreme Value	Weibull	Generalized Pareto	Fatique Life	Johnson SB	Gamma	Generalized Extreme Value	Dagum
Parameters	k, σ, μ	α, β, γ	α, β, γ	k, σ, μ	α, β, γ	k, σ, μ	α, β, γ	γ, δ, λ ξ	α, β, γ	k, σ, μ	k,α, β, γ
Maximum	0.35	0.211	0.03	0.3	0.39	0.03	0.27	0.012	0.50	1.45	0.60
Minimum	0.098	0.004	0.004	0	0	0	0.01	0	0.17	0.37	0.21
Mean	0.2	0.02	0.01	0.07	0.06	0.008	0.04	0.004	0.26	0.85	0.41
Standard deviation	0.05	0.03	0.007	0.06	0.06	0.007	0.04	0.003	0.07	0.23	0.07
Coefficient of Variation (CV)	19%	168%	56%	82%	109%	89%	89%	75%	27%	28%	19%
Skewness	Left	Right	Right	Right	Right	Right	Right	Right	Right	Right	Right
Validation of Control Chart (± 3 σ)	**100 %**	**97 %**	**100 %**	**98%**	**97%**	**100%**	**97%**	**100%**	**97%**	**100%**	**100%**

Moreover, it shows that chemical compositions among samples is not normally distributed which indicates variability among manufacturers row material used for producing rebar's. It is evident that, for most bars sizes of A 615 Grades 60, the mean for the chemical composition is not situated at the midpoint of the data range, indicating non-normal distributions.

4. Conclusion, recommendations and benefits

Chemical composition tests conducted on 68 samples of locally manufactured steel reinforcing rebar's in Saudi Arabia. The analysis of the steel bars from these manufacturers possessed percentage average Carbon, Phosphors, Sulphuer, Copper, Chromium, Manganese and Carbon Equivalent contents of 0.26, 0.02, 0.013, 0.079, 0.061, 0.852 and 0.402, respectively. Also it is shown shown that less than 3% of the steel failed to meet minimum ASTM standards for chemical composition. It is evident that, for most bars sizes of A 615 Grades 60, the mean for the chemical composition is not situated at the midpoint of the data range, indicating non-normal distributions, therefore, the characteristic chemical composition of locally produced steel bars is not consistent. The following recommendation and benefits can be deduced from this study.

1. The results are very important for the long term financial stability and mechanical viability and structural safety for this sector in KSA
2. Saudi Arabia Standard organization board should make sure, the standard given by them is strictly observed by all the local producers.

5. References:

[1] ASTM A 615/M615 -15. Standard Specification for Deformed and Plain Carbon-Steel Bars for Concrete Reinforcement. ASTM standards, Philadelphia: American Society for Testing and Materials.

[2] C. B. Prabir, P. Shylamoni and A. D. Roshan, "Characterization of steel reinforcement for RC structures: An overview and related issues" Jan. 2004, The Indian Concrete Journal page 19-30

[3] F. K. Hajeri-Al, Presentation on carbon content in steel rebar's, Saudi Arabia Standard Organization. 2015.

[4] J.L. Morán-López, , J. M. Sánchez, "Advanced Topics in Materials Science and Engineering" Springer Science & Business Media, 1993.

[5] F Djavanroodi, A Salman. "Variability of Mechanical Properties and Weight for Reinforcing Bar Produced in Saudi Arabia". IOP Conference Series: Materials Science and Engineering, Volume 230, Issue 1, Pages 012002

[6] F Djavanroodi, B Omranpour, M Sedighi [2012]. "Artificial neural network modeling of ECAP process". Materials and Manufacturing Processes, 28 (3), 276-281.

[7] C.K. Kankam, M. Adom-Asamoah. "Shear strength of concrete beams reinforced with steel bars milled from scrap metals". J Mater Des 2006;27:928–34.

[8] MH Shaeri, M Shaeri, MT Salehi, SH Seyyedein, F Djavanroodi [2015]. "Microstructure and texture evolution of Al-7075 alloy processed by equal channel angular pressing". Journal of Transactions of Nonferrous Metals Society of China Volume 25 Issue 5 Pages 1367-1375.

[9] D. E. Allen,"Statistical Study of the Mechanical Properties of Reinforcing Bars," Building Res. Note No. 85, Division of Building Res., National Res. Council, Ottawa, Canada, Apr., 8 pp.

[10] S. A. Mirza and J. G. MacGregor, "Variability of Mechanical Properties of Reinforcing Bars," Journal of the Structural Division, ASCE, Vol. 105, May, pp. 921-937, 1979.

[11] A. S.Nowak, and M. M. Szerszen, "Calibration of Design Code for Building (ACI 318): Part 1-Statistical Models for Resistance," ACI Structural Journal, Vol. 100, May-June, pp. 377-382, 2003.

[12] M. Bournonville, J. Dahnke, and D. Darwin, , "Statistical Analysis of The Mechanical Properties and Weight of Reinforcing bars" Structural Eng. and Eng. Materials, SL Report 04-1, 2004

[13] M. U. Jibrin, S. P. Ejeh, "Chemical Composition of Reinforcing Steel Bars in the Nigerian Construction Industry" Int. J. of Civil & Environmental Eng. IJCEE-IJENS Vol:13 No:01, Feb. 2013

[14] C.P. Joshi, and R. Ranganathan, "Variations in strength of reinforcing steel bars," Journal of the Institution of Engineers (India): Civil Engineering Division, Vol. 68, No. 6, 1988,pp. 309-312.

[15] S. Akyz, and M.Uyan, "Study on the reinforcing steel bars used in Turkey," TeknikDergi/Technical Journal of Turkish Chamber of Civil Engineers, Vol. 3, No. 1, 1992, pp. 136-138.
[16] A.M. Arafah, "Statistics for concrete and steel quality in Saudi Arabia," Magazine of Concrete Research, Vol. 49, No. 180, 1997, pp. 185-193.
[17] C. Galasso, E. Cosenza, and G. Maddaloni, "Statistical analysis of reinforcing steel properties for seismic design of RC structures," Proc. of the 14th European Conf. on Earthquake Eng, Ohrid, Macedonia, 2010, pp. 3715-3722.
[18] W. T. Lim, "Statistical Analysis of Reinforcing Steel Properties," Master's Thesis, University of Canterbury, Christchurch, New Zealand, June 1991, 151 pp.
[19] M. U. JIBRIN, S. P. EJEH, "Chemical Composition of Reinforcing Steel Bars in the Nigerian Construction Industry" International J. of Civil & Environmental Eng. IJCEE-IJENS Vol:13 No:01.
[20] http://www.metalsteelsaudi.com/?page_id=17
[21] ASTM A706/A706M - 2014, Standard Specification for Deformed and Plain Low-Alloy Steel Bars for Concrete Reinforcement. ASTM standards, Philadelphia: ASTM.

Acknowledgment:

This research was supported by Prince Mohammad Bin Fahad University. The support is greatly acknowledged. We are also thankful to Dr. M. M. Al Motari at Saudi Arabia Standard Organization (SASO), Dr K. F. Al-Hajeri, Mr M. Yaghoub at SABIC and Mr. H. M. Dowla at Al Ittifaq steel for providing their laboratories and assistance for the project. Their support is greatly acknowledged. Finally, we thank I. Babbili1; I. Ansari1; R. Al-Roqaiti1; O. Qassar2; N. Qahtani1; S. Aladhyani1; A. Alunaiz1; M. Almatrook1; M. Najar3, A. Asiz1 for their valuable input and assistance.

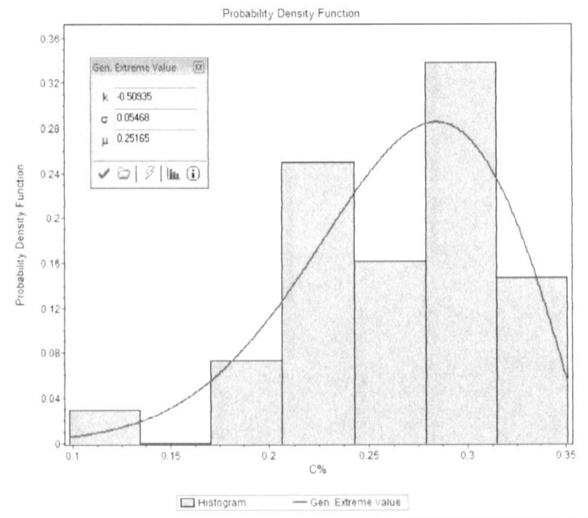

(a) Carbon (C): Generalized Extreme Value

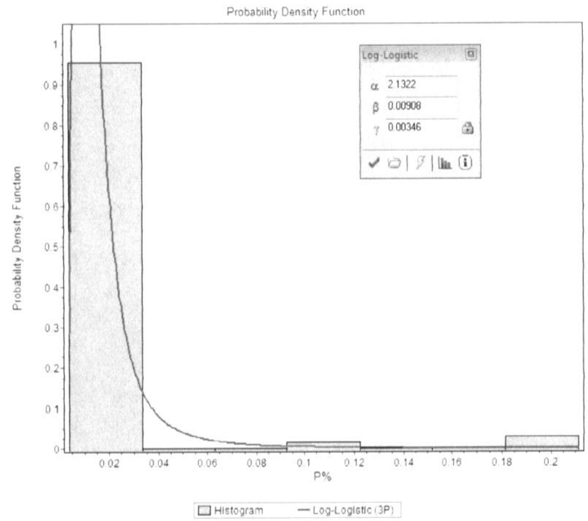

(b) Phosphorus (P): Log-Logistic

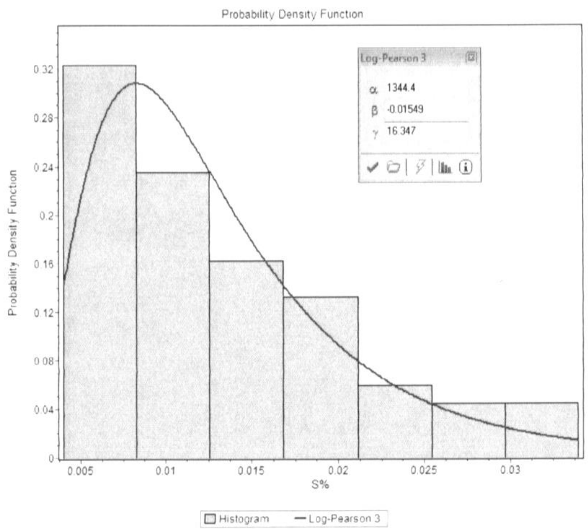

(c) Sulphur (S): Log-Pearson

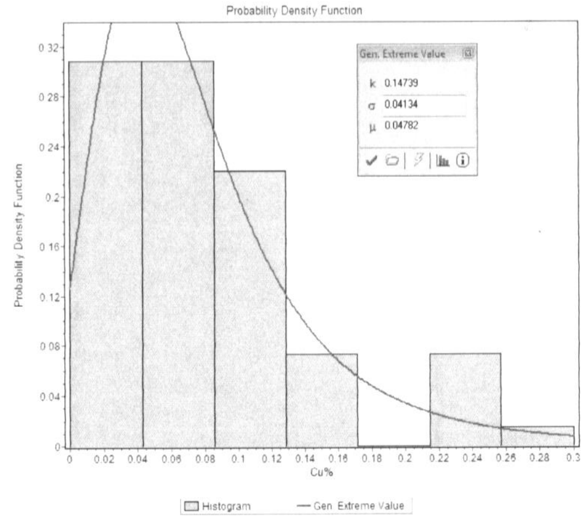

(d) Copper (Cu): Generalized Extreme Value

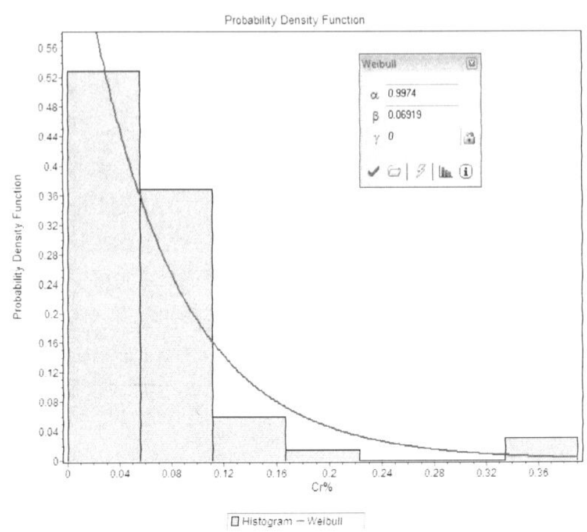

(e) Chrome (Cr): Weibull

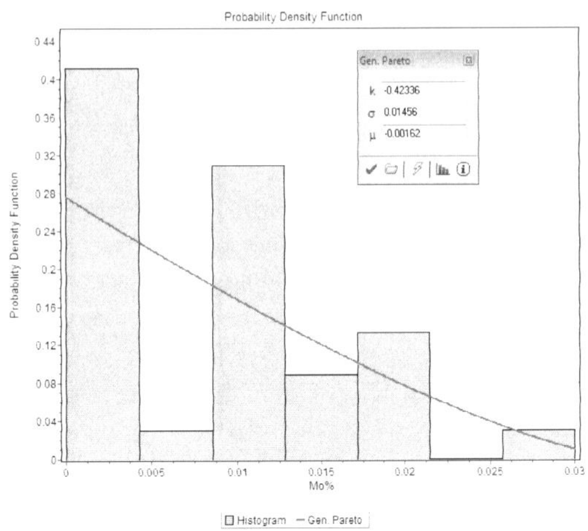

(f) Molybdenum (Mo): Generalized Pareto

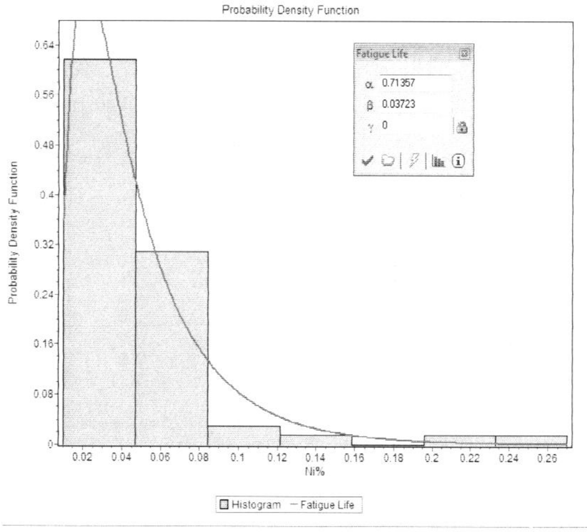

(g) Nickel (Ni): Fatique Life

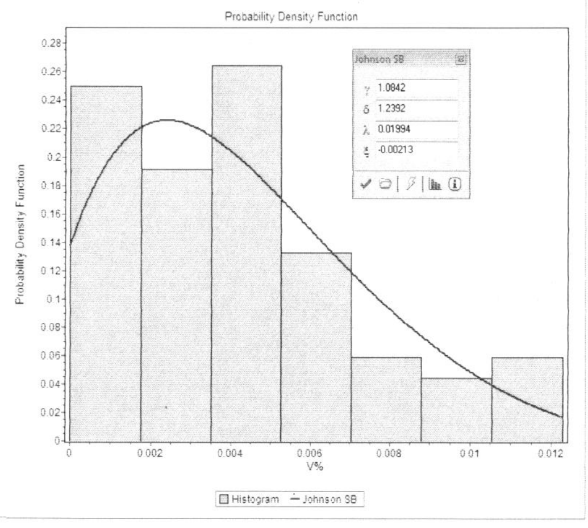

(h) Vanadium (V): Johnson SB

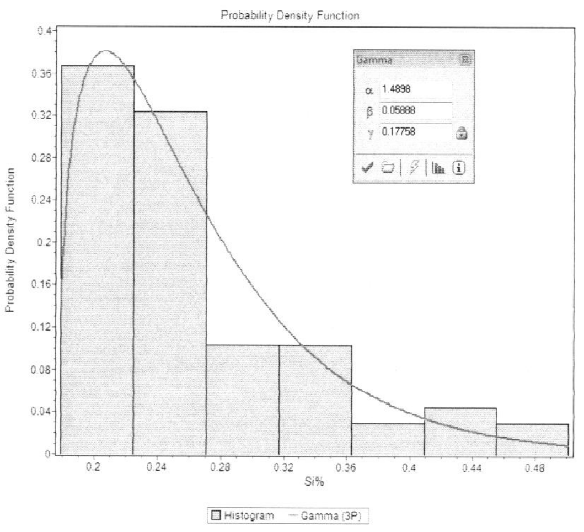

(i) Silicon (Si): Gamma

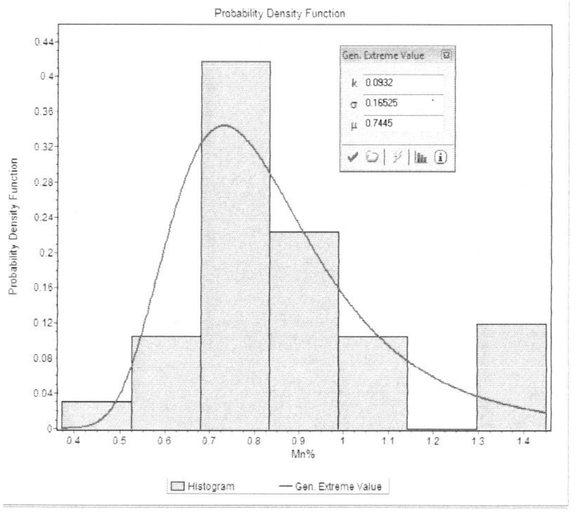

(j) Manganese (Mn) : Generalized Extreme Value

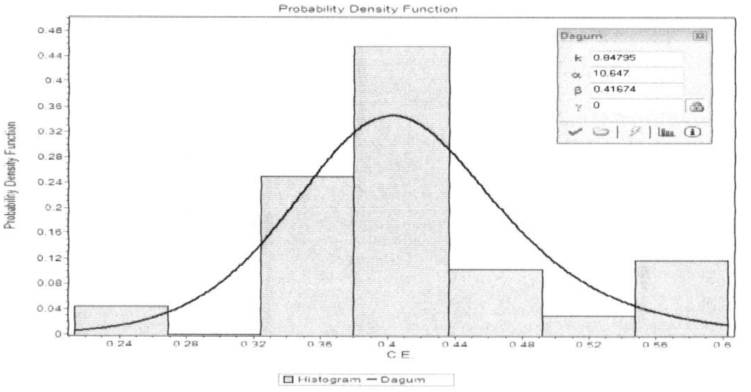

(k) Carbon Equivalent (C.E.): Dagum

Figure 2. Statistical analysis of chemical compositions

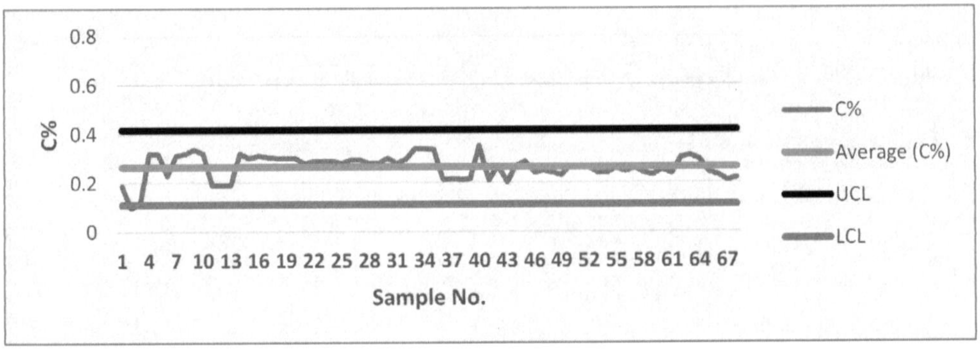

(a) Carbon (C): 100%

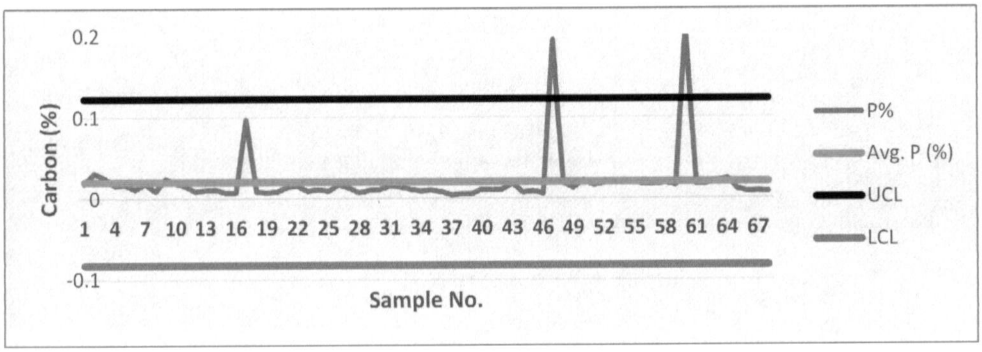

(b) Phosphorus (P): 97%

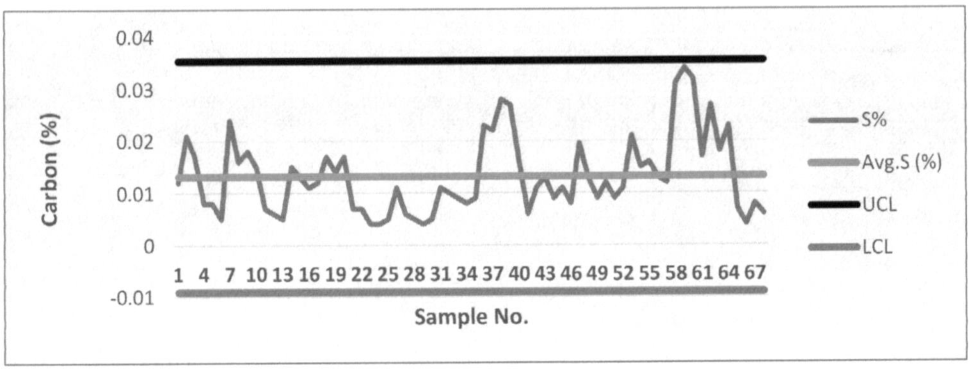

(c) Sulphur (S): 100%

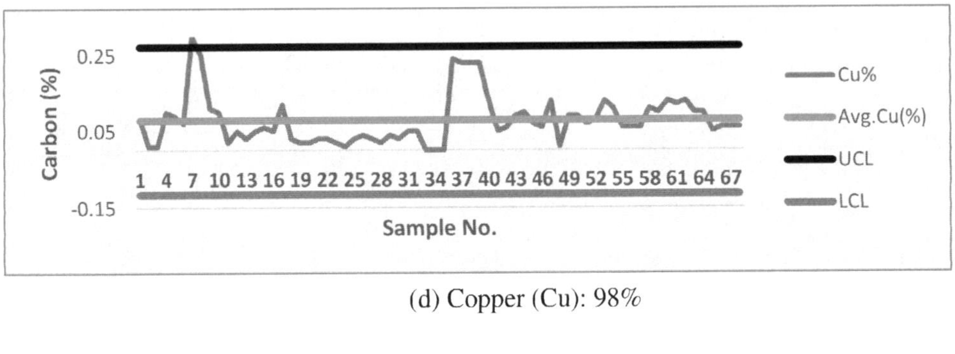

(d) Copper (Cu): 98%

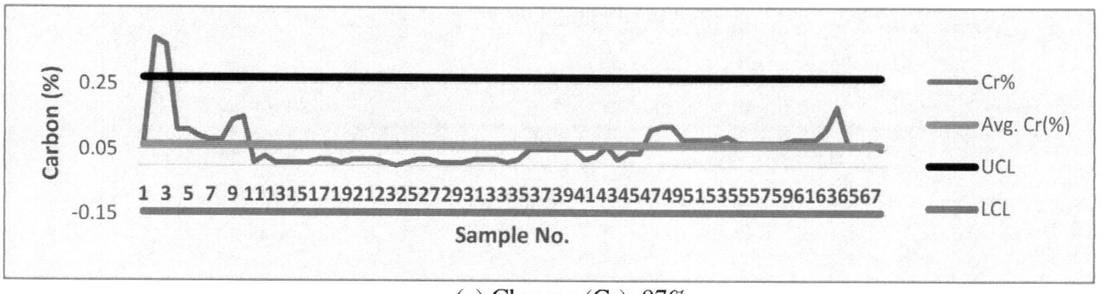

(e) Chrome (Cr): 97%

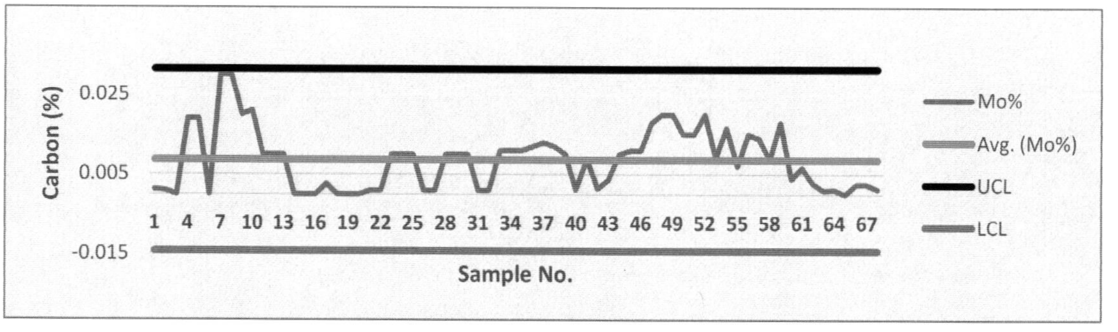

(f) Molybdenum (Mo): 100%

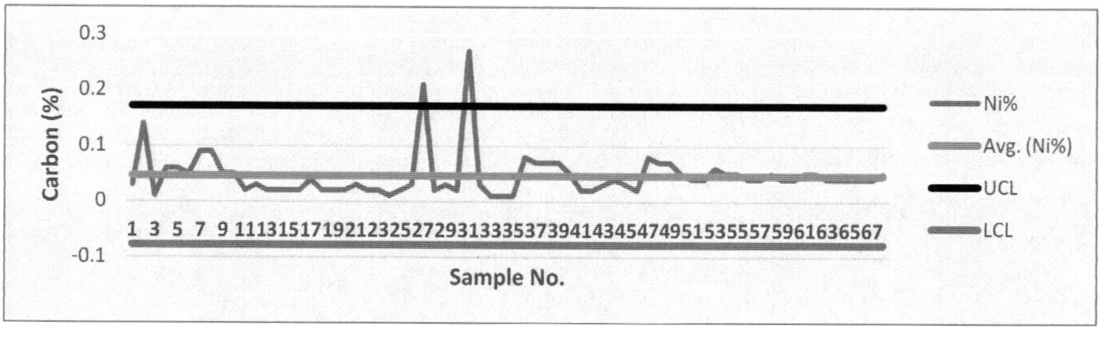

(g) Nickel (Ni): 97%

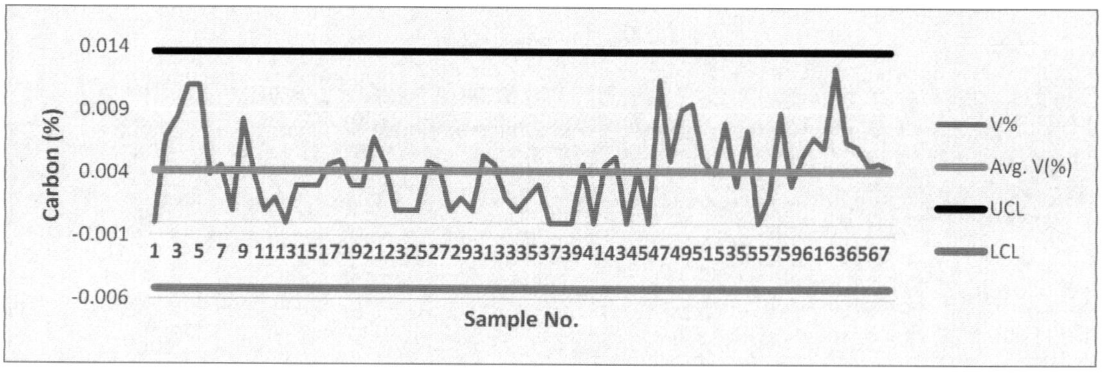

(h) Vanadium (V): 100%

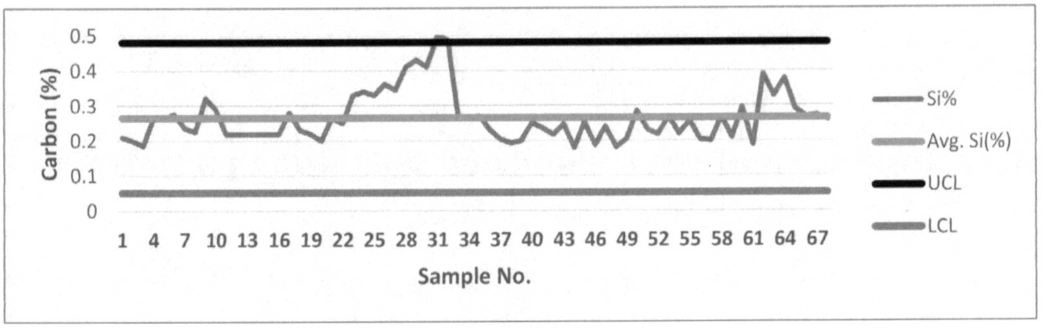

(i) Silicon (Si): 97%

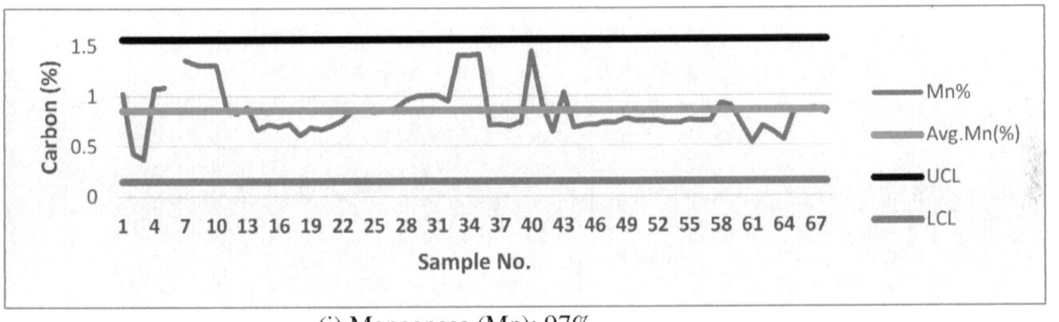

(i) Manganese (Mn): 97%

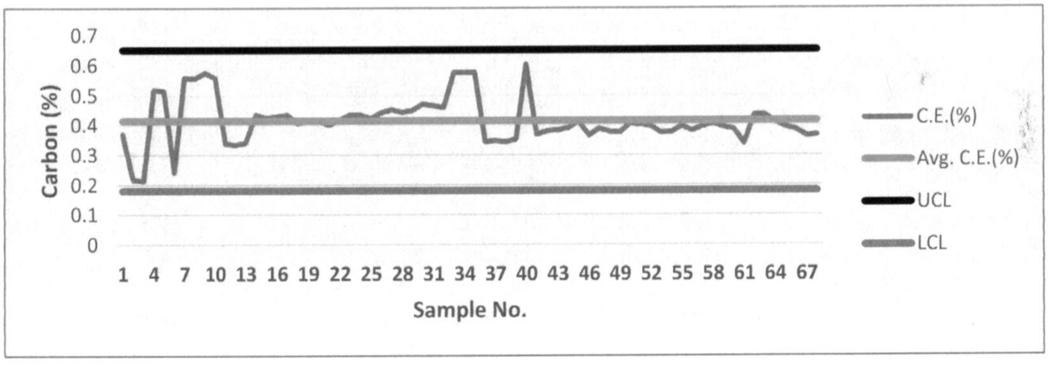

(j) Carbon Equivalent (C.E.) : 100%

Figure 3. Control charts of chemical compositions

IOP Conf. Series: Materials Science and Engineering **348** (2018) 012016 doi:10.1088/1757-899X/348/1/012016

The transformation of weak saturated soils using piles-drains for improving its mechanical properties

Z G Ter-Martirosyan[1], A Z Ter-Martirosyan[2] and V V Sidorov[3]

1 Main scientific researcher of Research and Educational Centre «Geotechnics», Moscow State University of Civil Engineering, 26 Yaroslavskoe shosse, Moscow, 129337, Russia
2 Head of Research and Educational Centre «Geotechnics», Moscow State University of Civil Engineering, 26 Yaroslavskoe shosse, Moscow, 129337, Russia
3 Scientific researcher of Research and Educational Centre «Geotechnics», Moscow State University of Civil Engineering, 26 Yaroslavskoe shosse, Moscow, 129337, Russia
E-mail: vitsid@mail.ru

Abstract. In practice of increased responsibility structures design there are often weak saturated clayey soils with low characteristics of deformability and strength take place on the construction site. In these cases, foundations using piles-drains of sandy or coarse material are recommended by norms, which is able to bear the load and to accelerate the consolidation process. The presented solutions include an analytical solution of the interaction problem between piles and slab raft foundation with the surrounding soil of the base with the possibility of extension of pile shaft. The closed-form solutions to determine the stresses in pile shaft and in the soil under the foundation slab are obtained. The article presents the results of large scale tests in the pilot area construction of major energy facilities in Russia.

1. Introduction

Availability of large capacity of the weak soil at the base greatly complicates its strengthening through complete replacement for sand or ground fill due to the large required depth of their laying and the enormous costs of the materials and time for its quality compaction (ensure the design quality parameters of the fill – the density of dry soil or compaction ratio of the entire thickness of the fill).
Currently, for pre-compaction of weak water-saturated bases, there are two ways: surface compaction using sand dams and embankments and deep compaction with the application of deep ramming, that creates substantial radial stresses in a weak base [1].
The proposed technique is based on the long-term consolidation of cohesive soils under permanent distributed load in the presence of the drains, which are columns made of crushed stone. After placing the columns on top of them they often constructs the horizontal gravel or sand fill, designed to collect and move water from the drains. The fill helps to reduce the non-uniform deformation of the base compressing as you remove water from it. This process is called the consolidation of the base [2].
In this methodology it is proposed to realize the compaction of the thick soft soil layer using gravel piles-drains.
1. The construction of the crushed stone piles is carried out by layer-by-layer compression of the material into the initial borehole, that leads to a large radial stress and extension of the walls of the initial borehole. Then in water-saturated soil around the column occurs the excess pore pressure that initiates the process of filtration consolidation.

2. Columns made of crushed stone have higher water permeability than the surrounding weak soil by several orders of magnitude, so they play the role of drains, which greatly speeds up the consolidation process, increasing the speed of the process of squeezing the water out of the ground and reducing the path of the horizontal filtration.
3. After making columns of crushed stone and the end of the consolidation process, the transformed composite (consisting of the piles-drains and the surrounding compacted soil) the base will work as a single converted massif on which it is possible to determine aggregate deformation and strength characteristics for the calculation of limit states.
For the construction of piles-drains an equipment for forcing the initial borehole is required and the equipment for pressing the crushed stone material into this borehole. For these purposes can be used any standard drilling equipment available.
The initial borehole is formed by pressing a working tool to the required depth. Thus, there is the first compaction step. The tool as usually is not removed. Then the bunker install on the tool and poured with crushed stone. After that, open the gate, and the crushed stone reaches the lower end of the tool. After the spill of crushed stone instrument raise to a predetermined height, and crushed stone, reached the bottom of the well, remains at the same depth and compacted by the pressing tool at a certain distance. Thus, the initial borehole is greatly expanded, the crushed stone pile-drain is formed, and the soil around the column is compacted in the radial direction. The compaction of surrounding soil causes by intensification of the consolidation process due to the increasing of excess pore pressure.
The most simple operating tool of the machine can be of a tube of appropriate rigidity (the ability to perceive vertical load of up to 300 t) with a closed bottom end [3].

2. Analytical solutions

Construction process of piles-drains.
For the design of weak soil base improvement with piles- drains, we need to consider two successive process - the interaction of surrounding soil with the expanding shaft (during construction) and the bearing process of the pile shaft and surrounding soil under the load as a part of the foundation [4]. The initial parameter as usual is the radius of piles, i.e. the radial movement of the pile shaft is a predetermined value. The outer limit of the work cell [fig.1] is determined by the arrangement of piles-drains, from which the degree of conversion of a weak base depends.

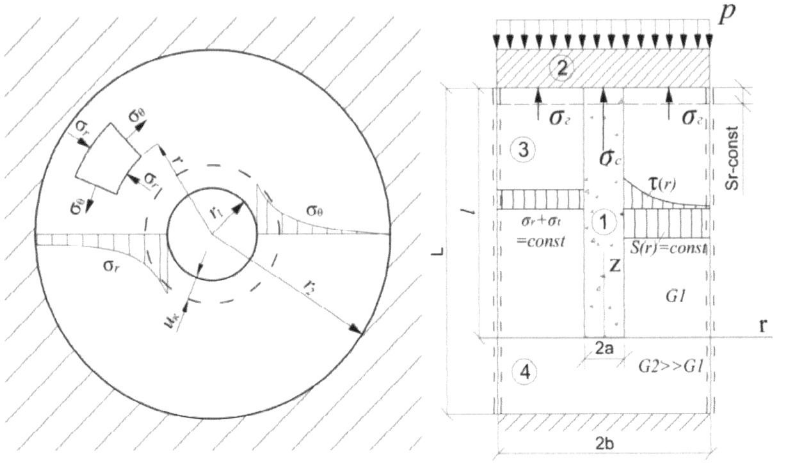

Figure 1. Design scheme of the initial borehole (left) and of the transformed base work cell (right).

Based on the joint review of geometrical, physical and balance equations, the problem of stress-strain state of the soil cylinder can be reduced to the solution of differential equations [5] in displacements of the form:

$$\frac{d^2u}{dr^2}+\frac{1}{r}\cdot\frac{du}{dr}=\frac{u}{r^2} \tag{1}$$

You can then record the resulting expression for finding the radial and tangential stresses on the pile-soil contact.

$$\sigma_r=\frac{E_2A}{(1+\nu_2)\cdot\nu_2^{/}}\left(1+\frac{r_2^2}{r^2}+\left(1-\frac{r_2^2}{r^2}\right)\xi_2\right)+\frac{\sigma_{z2}\cdot\xi_2}{(1+\nu_2)\cdot\nu_2^{/}} \tag{2}$$

$$\sigma_\theta=\frac{E_2\cdot A}{(1+\nu_2)\cdot\nu_2^{/}}\left(\left(1-\frac{r_2^2}{r^2}\right)+\xi_2\left(1+\frac{r_2^2}{r^2}\right)\right)+\frac{\sigma_{z2}\cdot\xi_2}{(1+\nu_2)\cdot\nu_2^{/}} \tag{3}$$

To determine the vertical load to be transferred by ramming equipment on the ground, you need to use the equation of limit equilibrium, where the main stress will be present as the value of vertical and radial stresses [6]. This vertical stress should be a limit to the material layer to destroy and force it to take a design position in the borehole that has already expanded well. Limit equilibrium equation can be represented as follows [7]:

$$\frac{\sigma_z^*-\sigma_r}{\sigma_z^*+\sigma_r+2c\cdot ctg\varphi}=\sin\varphi \tag{4}$$

Then the value of the limit vertical stress is expressed as follows:

$$\sigma_z^*=\frac{\sigma_r(1+\sin\varphi)+2c\cdot\cos\varphi}{(1-\sin\varphi)} \tag{5}$$

Based on (4) we can express the value of N required to grow the initial borehole to desired size:

$$N^*=\pi\cdot r_1^2\cdot\sigma_z^* \tag{6}$$

The interaction between soil pile of finite stiffness of the surrounding soil in the foundation, taking into account the extension of its diameter in the process of loading [8].
Considering the equality of radial displacements at the pile radius ($r=r_l$) and taking into account the equality of the deformation, we can obtain the expression relating the load on the foundation plate p and of the resulting lateral expansion of the pile under load u_l. The expression is complicated and represents simply with the help of special coefficients, which are presented in as follows:

$$u_1=p\left(\frac{G}{F}+\frac{J}{K}\right):\left(\frac{I}{K}+\frac{H}{F}\right) \tag{7}$$

The coefficients A,B,C,D,E,F,G,H,J,K depend on the geometrical and mechanical parameters of the pile-drains and the surrounding soil around it and under its tip (Ter-Martirosyan Z.G. et.al., 2016) [fig.2].

Definition of the meaning given value of the deformation modulus of the soil base compacted with crushed stone piles-drains.
After construction of the piles-drains on the site and consolidation of the weak base, it is necessary to consider the work of composite base under load. The main question at this stage is finding the settlement of the base. As bearing elements of this system are the piles and the surrounding soil and to consider them separately is very difficult. Therefore the equation for obtaining the deformation modulus is required.

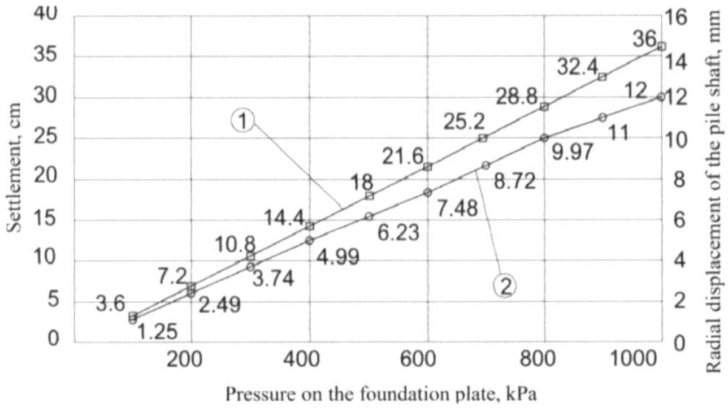

Figure 2. Pressure-settlement and pressure-radial displacements curves.

The dependences for definition of the deformation modulus in cases where under the tip of the pile-drain lies dense soils and punching of the soil under the tip is almost impossible, and also for the case when the soils under tip are deformable and punching is possible. In the first case the expression has the form:

$$\overline{m} = \frac{m_c \cdot m_s}{m_s \cdot \lambda + m_c(1-\lambda)} \tag{8}$$

where $\overline{m}$ - relative compression coefficient of the entire work cell;
m_c - relative compression coefficient of the pile material;
m_s - relative compression coefficient of the surrounding improved soil of the base.

$\lambda = \frac{a^2}{b^2}$ - volume deformation of the extending pie during the construction process, a - final radius of the pile, b - radius of the work cell.

Values of relative compression coefficient of the soil and pile-drain material can be obtained like this:

$$m = \frac{\beta}{E} \tag{9}$$

where β is a coefficient of the lateral pressure and E is a deformation modulus.

In case of possibility of soil punching under the tip of the pile-drains to obtain a more complex dependency that is more convenient to write in short form:

$$\overline{m} = \frac{D \cdot E}{l} \tag{10}$$

where, $D = \frac{m_s \cdot l \cdot \omega}{B + A \cdot \omega \cdot K_0 \cdot C}$, $E = \frac{e^{\omega l} - e^{-\omega l}}{\omega E_c\left(e^{\omega l} + e^{-\omega l}\right)} + \frac{A \cdot K_0 \cdot 2}{\left(e^{\omega l} + e^{-\omega l}\right)}$

$$B = \frac{1}{E_c} \cdot \frac{e^{\omega l} - e^{-\omega l}}{e^{\omega l} + e^{-\omega l}} + \frac{m_s \cdot \lambda}{1-\lambda} \cdot \frac{e^{\omega l} - e^{-\omega l}}{e^{\omega l} + e^{-\omega l}}, \quad C = \frac{2}{e^{\omega l} + e^{-\omega l}} + \frac{m_s \cdot \lambda}{1-\lambda} \cdot \frac{\left(2 - e^{\omega l} - e^{-\omega l}\right)}{e^{\omega l} + e^{-\omega l}}$$

$$A = \frac{2}{e^{\omega l} + e^{-\omega l} + E_c \cdot K_0 \cdot \omega\left(e^{-\omega l} - e^{\omega l}\right)}, \quad K_0 = \frac{\pi a(1-v_2)}{4G_2}, \quad \omega^2 = \frac{6G_1}{a(b-a)E_c}$$

l - height of the pile-drain, G_l - shear modulus of the soil.

Expression (10) is more complicated for engineers but can be used more effective with help of any mathematical computional complex like MathCAD, MathLAB etc.

3. Field and laboratory soil and piles-drains tests

For reliable design of major facilities of nuclear power had created an 50x15m experimental area on which were constructed three sections of compacted base with the pitch of the piles of 1 m, 1.2 m and 1.5 m, of initial diameter 325 mm, 426 mm and 530 mm, respectively. After construction of the compacted piles and completing the consolidation process in the base, the field load tests [9] of the pile material were carried out [fig.3].

Table 1. Summary results of field testing of the piles-drains material

Depth,m	Soil	Deformation modulus E, MPa	Average value of E, MPa
7.0	crushed stone	120.5	140.5
7.0		106.9	
7.0		104.3	
7.0		146.7	
7.0		158.2	
7.0		206.6	

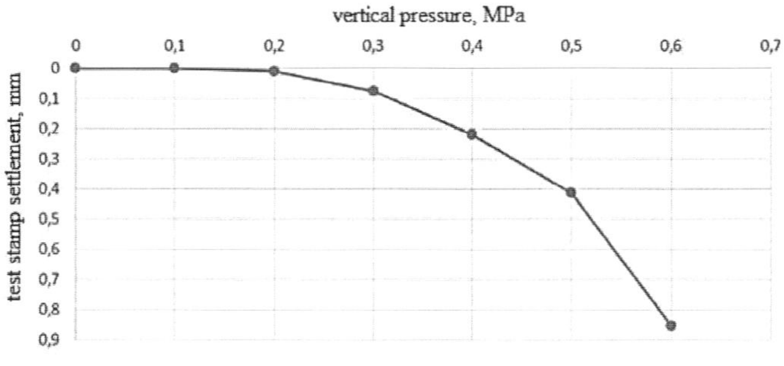

Figure 3. Load-settlement curve of the test flat stamp.

Figure 4. Triaxial test machine for coarse soil material and non-homogenous soils with large size inclusions.

Tests have shown that crushed-stone material is sufficiently strongly compacted in the process of piling according to the considered technology [10]. Getting high deformation characteristics of the material of the columns will lead to a significant increase in the total given deformation characteristics of compacted base as a whole, as the columns and compacted soil around them will work together [table 1].

To determine the deformation characteristics of the compacted base as a whole (composite base consisting of gravel piles and compacted soil around it) MSUCE specialists carried out testing of the work cell with flat stamp with an area of 7854 cm2 (diameter 1000 mm) in the well and at the bottom of the pit at a depth of 6.5 m, the absolute mark of 145.00 [fig. 5,6]. Just done seven tests. The tests were carried out in accordance with GOST 20276-2012 "Soils. Field methods for determining the strength and deformability". The resulting values of the deformation modules was 45.07 MPa, 42.88 MPa , 56.64 MPa, 44.09 MPa, 37.14 MPa, 38.87 MPa and 42.01 MPa, which generally exceeds the design value of 40 MPa. Tests have confirmed the assumptions about the joint work of members of composite base, arranged according to the presented technology, as well as the correctness of the statements themselves to determine the given deformation modulus. Were also conducted tests on lab equipment [fig.4] obtained similar results with field studies.

Figure 5. Loading flat stamp of diameter 1000 mm (left) and general view of the test equipment located in the pit (right).

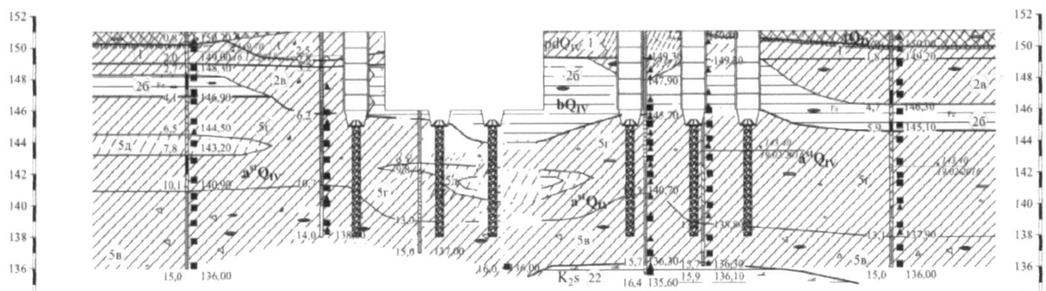

Figure 6. Geology cross section of the experimental area showing the locations of test piles-drains.

4. Conclusions

1. The technology for the hardening of weak bases is one of the most cost-effective and efficient. Low cost is provided by the simplicity of the technological cycle, requires simple machines and mechanisms for its implementation.
2. Created analytical solutions are sufficient to determine the technological parameters of the compaction - pitch of the pile, the ultimate diameter of the working stroke of the equipment and given deformation characteristics of the transformed base.
3. Conducted on the experimental area tests of the piles-drains material, the working cell of the transformed base and surrounding compacted soil showed the efficiency of the technology and sufficient approximation of the values of mechanical parameters of soils to design parameters.

4. Thus, the technique of compacting base with crushed stone piles-drains can compete with traditional methods of compaction and with a more expensive use of deep foundations.

5. References

[1] SP 24.13330.2011. Pile foundations. FAU FCS (2011).
[2] SP 22.13330.2011. Soil bases of buildings and structures. FAU FCS (2011).
[3] Ter-Martirosyan A Z, Ter-Martirosyan Z G and Sidorov V V 2016 Interaction of compaction piles with surrounding soil with due regard for pile diameter expansion. *Soil Mech. and Found. Eng. Iss.3* 53 166-73.
[4] Bezvolev S G 2011 Problems of design and analysis of foundations when large groups of piles and other vertical elements to transform the soil massif. *Geotechnics* **3** 30-67.
[5] Bezukhov N I 1968 *Bases of Theory of Elasticity, Plasticity and Creep* (Moscow: High school).
[6] Ter-Martirosyan Z G, Sidorov V V and Strunin P V 2013 Calculation of the stress-strain state of a single compressible barrets and piles interactied with the soil. *Civ. Engineering* **9** 18-22.
[7] Timoshenko S P, Gudier J 1975 *Theory of Elasticity* (Moscow: Science).
[8] Grevcev A A 2012 The theory of expansion of the cavity and limiting resistance of the soil under the toe of driven piles in sandy soils. *Civ. Engineering* **9** 2-5.
[9] Barvashov V A 2009 Experimental and theoretical studies of computational models of slab-pile foundations. *Soil Mech. and Found. Eng.* **5** 32-9.
[10] Ter-Martirosyan Z G and Abdulmalek A S 2007 The stress-strain state of the compacted base. *Soil Mech. and Found. Eng.* **6** 8-11.

Acknowledgments

This work was financially supported by Ministry of Education and Science of the Russian Federation (1.4984.2017/6.7). All tests were carried out using research equipment of The Head Regional Shared Research Facilities of the Moscow State University of Civil Engineering.

IOP Conf. Series: Materials Science and Engineering **348** (2018) 012017 doi:10.1088/1757-899X/348/1/012017

Study on the influence of design parameters on the damping property of glass fiber reinforced epoxy composite

A Bhattacharjee[1] and B.K. Nanda[2]

1 Department of Mechanical Engineering, National Institute of Technology, Rourkela, Odisha, India
2 Professor, Department of Mechanical Engineering, National Institute of Technology, Rourkela, Odisha, India
E-mail: 515ME1003@nitrkl.ac.in

Abstract. Fiber reinforced composites are widely used in industrial applications due to their high strength, light weight and ease in manufacturing. In applications such as automotive, aerospace and structural parts, the components are subjected to unwanted vibrations which reduce their service life, accuracy as well as increases noise. Therefore, it is essential to avoid the detrimental effects of vibrations by enhancing their damping characteristics. The current research deals with estimating the damping properties of Glass fiber reinforced epoxy (GFRE) composites. Processing of the GFRE composites is carried out using hand-lay technique. Various design parameters such as number of glass fiber layers, orientation of fibers and weight ratio are varied while manufacturing GFRE composites. The effects of variation of these design parameters on damping property of GFRE composites are studied extensively.

1. Introduction

The use of fiber reinforced polymer (FRP) matrix composites is increasing due to inherent advantages of high strength, easy manufacturing techniques, light weight, high damping, etc. These found wide industrial applications such as automotive, aerospace, marine, medical, sports, machine tools and various other structures [1]. Further, polymer matrix composites can be produced in any complex shape requiring less energy. Examples of such polymers are epoxy, polyester, urethane, etc. However, the most commonly used polymer is epoxy which is usually reinforced with glass fibers and is specifically used for structural applications. In such applications it is essential to know the damping property of the composites.

Chandra et al. [2, 3] have mentioned in their work that damping in composite materials is different from that of conventionally used metals. They reported that the various ways in which energy is dissipated from FRP composites are-viscoelastic behaviour of matrix or fibers, interphase damping, visco-plastic damping and thermo-elastic damping. They further used an integrated finite element method or strain energy approach to determine the loss factor for two and three phase composite with interfacial discontinuity. They have also carried out research on the theory of damping mechanism of composite materials such as micromechanical, macro-mechanical and viscoelastic. Srinivas et al. [4] have studied the mechanical and chemical properties of various natural fiber reinforced composites such as jute, hemp, kenaf, flax, etc. and compared the results with synthetic fibers. Suresha et al. [5] have demonstrated that filling glass epoxy composite with graphite particulates improves the mechanical properties such as hardness, tensile strength, tensile modulus, percentage elongation as well as wear resistance. Deogonda et al. [6] have studied the effect of reinforcement of TiO_2 and ZnS particles on the mechanical properties of GFRE composites. They found that with increase in ZnS

filler content, the mechanical properties of composites increases compared to the other combination of composites. Biswas et al. [7] have shown that the erosion characteristics of glass and bamboo fiber epoxy composites improve when reinforced with red mud particulates.
Therefore, the current research focuses on studying the effect of various design parameters on the damping property of glass fiber reinforced epoxy (GFRE) composite. The various design parameters that are taken into account are fiber loading, fiber orientation and number of glass fiber layers.

2. Experimentation details

The experimental study comprises of estimating the damping properties of GFRE composite. Woven glass fiber mat is used with epoxy resin and hardener having trade-name Bisphenol-A-Diglycidyl ether and tri-ethylene tetra-amine respectively. The matrix material comprises of epoxy and hardener, mixed in the weight ratio of 10:1. The wet lay-up technique is used to fabricate the GFRE composites as the process is simple and inexpensive. This open mould process is suitable for fabricating any variety and size of polymer matrix composites. The process is easy compared to other manufacturing processes as it doesn't require any complex tools or instruments and can also be done at normal environmental conditions. Mould releasing agent such as sheet and spray are used to cover the surface of the mould. Care is taken to ensure thorough mixing of the epoxy and hardener as well as proper matrix coating on the glass fiber surface. The fiber surface is frequently rolled with a hand roller to remove any trapped gas bubble or tangled glass fibers. Each cast composite is first cured by applying a load of about 400 N for a period of 24 hours and then cured in air for 24 hours. Utmost care is taken to maintain homogenous distribution of fiber and matrix material and to avoid any defect in the final composite structure.
Test samples of 350×50×5 mm dimensions are cut from the cured GFRE samples using a diamond cutter. An experimental setup consisting of a heavy and rigid frame has been prepared to hold the fixed end of the cantilever beam as shown in Figure 1. In order to ensure negligible support damping, the composite beams are stiffly fixed to the frame in order to avoid any rotation and achieve rigid cantilever condition. Vibration exciter is placed below the free end of the cantilever beam to impart forced excitations to the beam. A contact type accelerometer is placed on the free end to sense the signals and supply it to the digital storage oscilloscope (DSO). Function generator is used to apply required loading frequency for excitations.

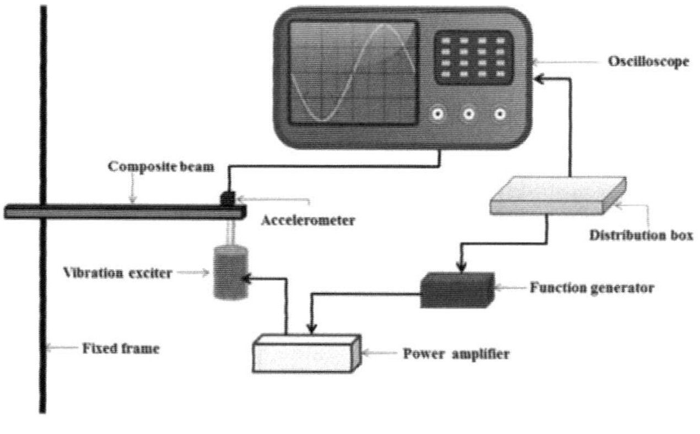

Figure 1. Schematic diagram of the experimental setup.

Logarithmic decrement method is used to calculate the damping parameters from amplitude versus time plot obtained from DSO. The following equations are used to calculate the damping ratio and natural frequency from the curves.

$$\delta = \frac{1}{n}\ln(\frac{x_1}{x_{n+1}}) \tag{1}$$

$$\zeta = \frac{\delta}{\sqrt{(4\pi^2 + \delta^2)}} \tag{2}$$

$$\omega_d = \frac{2\pi}{\Delta T} \tag{3}$$

$$\omega_n = \frac{\omega_d}{\sqrt{1-\zeta^2}} \tag{4}$$

where δ, ζ, ω_d, ω_n, x_1, x_{n+1}, ΔT and n denotes logarithmic decrement, damping ratio, damped natural frequency, natural frequency, peak amplitude of first cycle, peak amplitude of $(n + 1)^{th}$ cycle, time period between two successive amplitude peaks and number of cycles respectively. The current research focuses on the effect of various parameters such as fiber orientation, fiber loading and number of fiber layers on the damping property of GFRE composite.

3. Results and discussions

3.1. Effect of weight ratio or fiber loading

The ratio of weight of fibers reinforced to the weight of resin is known as fiber loading. The damping of epoxy matrix is quite significant compared to glass fibers. This is due to the visco-elastic nature of epoxy which helps in dissipating energy. Moreover, the interphase region adjacent to the fiber surface also contributes in energy dissipation. Figure 2 illustrates the SEM (scanning electron microscope) micrographs of GFRE composites of 7:5 weight ratio of glass fiber to epoxy at 500x magnification. In the figure, the interfacial region between the fiber and the matrix is indicated using red arrows.

Figure 2. SEM micrographs of GFRE composite with 7:5 weight ratio.

The influence of fiber loading on the damping behavior of GFRE composites is evaluated. For this purpose, distinctive weight ratio of 7:7, 7:6, 7:5 and 7:4 of glass fibers and blend of epoxy and hardener is used for fabricating GFRE composites. The number of glass fiber layers used for fabricating the GFRE composites is 12. For performing damping experiments, 300 mm cantilever length of GFRE is used with excitation frequency of 5 Hz. Amplitude vs. time plots are obtained from experimental analysis and logarithmic decrement method is used to find out the damping ratio and natural frequency. The damping ratio for each weight ratio is calculated and plotted as shown in Figure 3.

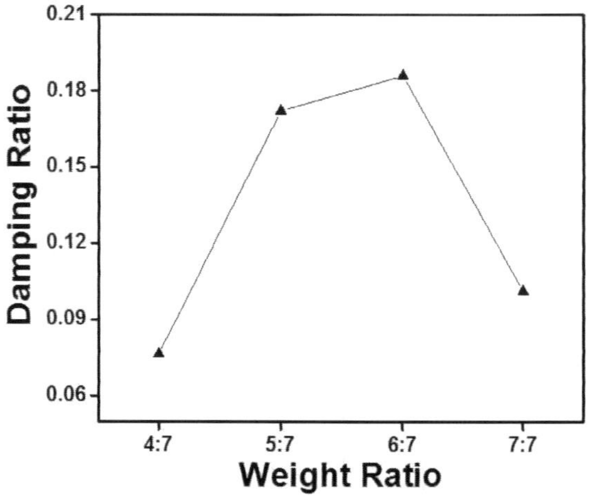

Figure 3. Plot of damping ratio versus weight ratio of GFRE composites.

The dissipation of energy in FRP composites occur either due to the viscoelastic nature of the matrix or due to the interfacial region between the fiber and the matrix. From *F*igure 3, it is evident that damping ratio increases with increase in the percentage of epoxy content up to 46% (approx.) i.e., 7:6 weight ratio and then decreases. The GFRE composite with 50 weight % of matrix has the least damping. The increase in damping of GFRE composites up to 46% of matrix is due to dissipation of energy through epoxy matrix. However, in case of 50% epoxy or fiber content, the damping is decreased due to reduced weight fraction of glass fibers as compared to other 3 types of GFRE composites. In this case, the damping due to fiber-matrix interface predominates the damping due to viscoelastic matrix.

3.2. Effect of woven fiber-mat orientation

A total of three samples were made for the experiment of damping. The materials used for making these samples are the glass fiber and epoxy resin with proper hardener. Each sample has 12 layers of the glass fiber. The weight of glass fiber was measured and according to that the weight of epoxy-hardener mixture was decided. As the weight ratio taken was 1:1, the epoxy-hardener weight is same as glass fiber. The epoxy and hardener were mixed in the ratio of 10:1.

The difference in these samples is the orientation of the glass fibers. The orientation has been changed while fabricating these GFRE composites. The configuration of these samples is:

- 0 – 90° orientation
- 45 – (-45)° orientation
- 0 – 90 – 45 – (-45) ° orientation

To measure damping, composite cantilever beams of length 300 mm are used and the tests are performed at 5 Hz loading frequency. Amplitude vs. time plots obtained from the digital oscilloscope is used to calculate the damping ratios of the three composite samples. The damping ratio for the three composite of different fiber orientations is calculated and is plotted in Figure 4.

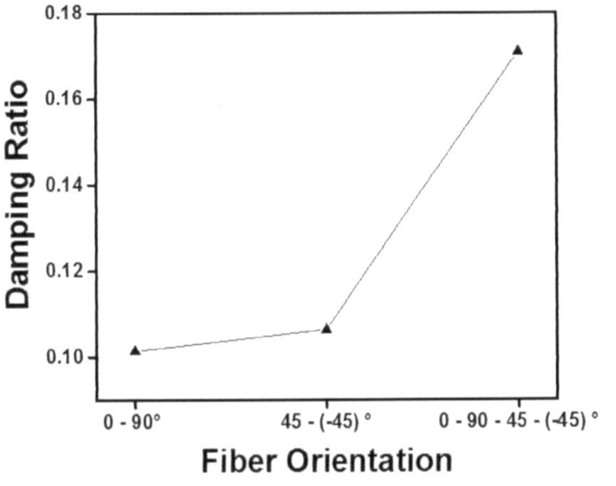

Figure 4. Plot of damping ratio versus fiber orientation of GFRE composites.

It can be seen from Figure 4 that the damping ratio is highest for 0–90–45–(-45) ° cross ply laminate. The angle orientation has lowered the damping value of 0–90° laminate compared to 45–(-45) ° orientation though, the difference is quite negligible. However, in case of 0–90–45–(-45) ° orientation, the energy dissipation is more due to cross-ply laminate arrangement which facilitates higher damping ratio.

3.3. Effect of number of fiber layers

The effect of number of fiber layers or the thickness of composite laminate on damping of composites is assessed by determining the damping ratio of samples prepared by stacking woven glass fiber layers at 0 - 90° orientation with distinct number of glass fiber layers as 9, 12 and 15. The weight ratio of 1:1 of glass fiber to the epoxy-hardener mixture is used. A cantilever length and excitation frequency of 300 mm and 5 Hz respectively is maintained. Logarithmic decrement method is used to calculate the damping ratio from amplitude vs. time plots obtained from experiments. The damping ratio of composite laminates for different number of glass fiber layers is shown in Figure 5.

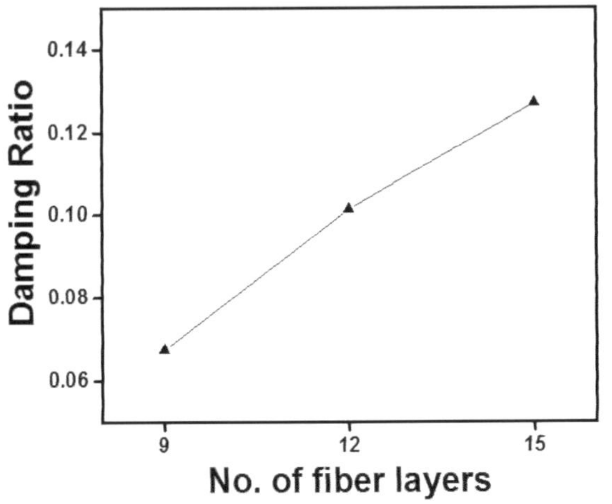

Figure 5. Plot of damping ratio versus number of fiber layers of GFRE composites.

It is seen that as the number of woven fiber layers increases, the damping ratio of the cantilever beam increases. The GFRE composite sample with 15 layers of woven fiber has maximum damping ratio. The increase in number of glass fiber layers facilitated the increase in dissipation of vibrational energy.

Moreover, the increased epoxy matrix content too enables better damping. Therefore, the damping ratios of the GFRE composites have increased with increasing fiber layers.

4. Conclusions

The experimental study of the effect of various design parameters of GFRE composites on the damping property leads to the following conclusions:

The damping ratio of GFRE composites increases with increase in weight of fiber matrix up to 46% and then found to decrease. The increase in damping is due to the dissipation of energy via the visco-elastic epoxy while the decrease is due to reduced weight fraction of glass fibers.

The damping ratio of GFRE composite with 0 – 90 – 45 – (-45) ° orientation has highest damping ratio compared to 45 - (-45) ° and 0 - (90) ° orientations due to cross-ply laminate arrangement enabling better energy dissipation.

The damping ratio of GFRE composites increases with increase in number of glass fiber layers.

5. References

[1] Kaw A K 2006 *Mechanics of Composite Materials* (London: CRC press Taylor and Francis group) p1-49

[2] Chandra R, Singh S P and Gupta K 2003 A study of damping in fiber-reinforced composites *J Sound Vib.* **262(3)** 475-96

[3] Chandra R, Singh S P and Gupta K 1999 Damping studies in fiber-reinforced composites- a review *Compos. Struct.* **46(1)** 41–51

[4] Srinivas K, Naidu A L and Raju M V A 2017 A review on chemical and mechanical properties of natural fiber reinforced polymer composites *Int. J. Perform. Engg.* **12(2)** 189-200

[5] Suresha B, Chandramohan G, Renukappa N M and Siddaramaiah 2006 Mechanical and tribological properties of glass-epoxy composites with and without graphite particulate filler *J. App. Polym. Sci.* **103** 2472-80

[6] Deogonda P and Chawla V N 2013 Mechanical property of glass fiber reinforcement epoxy composites *IJSER* **1(4)** 6-9

[7] Biswas S and Satapathy A 2010 A comparative study on erosion characteristics of red mud filled bamboo-epoxy and glass-epoxy composites *Mater. Design* **31** 1752-67

IOP Conf. Series: Materials Science and Engineering **348** (2018) 012018 doi:10.1088/1757-899X/348/1/012018

Plane waves in magneto-thermoelastic anisotropic medium based on (L-S) theory under the effect of Coriolis and centrifugal forces

Meshari Alesemi

1. Department of Mathematics, Faculty of Science, Jazan University, Jazan, Saudi Arabia
2. E-mail: m.alesem@jazanu.edu.sa

Abstract. The objective of this research is to illustrate the effectiveness of the thermal relaxation time based on the theory of Lord-Shulman (L-S), Coriolis and Centrifugal Forces on the reflection coefficients of plane waves in an anisotropic magneto-thermoelastic medium. Assuming the elastic medium is rotating with stable angular velocity and the imposed magnetic field is parallel to the boundary of the half-space. The basic equations of a transversely isotropic rotating magneto-thermoelastic medium are formulated according to thermoelasticity theory of Lord-Shulman (L-S). Next to that, getting the velocity equation which is illustrated to show existence of three quasi-plane waves propagating in the medium. The amplitude ratios coefficients of these plane waves have been given and then computed numerically and plotted graphically to demonstrate the influences of the rotation on the Zinc material.

1. Introduction

Recently, many results have been shown in the literature to the generalized thermo elasticity theories that depicted the second sound influence that allows the waves propagating with finite speed. Therefore, generalized theories were proved to shown more factual outcomes than the conventional models of thermoelasticity. The generalized thermoelasticity established on relaxation time effects was first presented by Muller [1]. Then, a more explicit theory where the temperature proportions are introduced through the constitutive relations was considered in the references [2] and [3]. Furthermore, Lord and Shulman's theory of generalized thermoelasticity utilizing one relaxation time expects the propagation of waves is finite [4]. These theories were developed by Dhaliwal and Sherief [5] to include the impacts of anisotropy. One may go through Hernarski and Ignaczak [6] for presenting and review of the generalized thermoelasticity.

The earliest works on the topic can exist in [7] and [8]. Abd-alla et al. [9], [10] and [11] displayed the phenomenon of the reflection and transmission or reflection only of waves propagating in a plane in the context of many assumptions under suitable boundary conditions. Othman and Song [12] and [13] have presented some problems in different hypotheses about magneto-thermo-elastic waves when the medium of the reflected wave is homogeneous and isotropic. Ting [14] investigated a surface wave propagation in an anisotropic medium under the effect of rotation.

Wave propagations in a generalized thermoelastic medium with additional variables such as magnetic field, anisotropy, microstructure, temperature and other variables supply essential details about the presence of new or changed waves. These details are beneficial for experimental seismologists in improvements earthquake estimation. Some relevant investigations on the reflection and transmission

wave propagation in a plane based on anisotropic the generalized thermoelastic media are considered by various authors such as Kumar and Singh [15] and Singh and Yadav [16].
Recently, Abd-alla et al. [17-21] studied several problems with regards to the reflections or the reflections and refractions phenomenon in anisotropic smart piezoelectric or in thermo-piezoelectric materials with the presence of the pre-stresses or not.

2. Formulation of the problem

Assuming a homogeneous and transversely isotropic elastic structure is rotating with steady angular velocity $\boldsymbol{\Omega} = \Omega \boldsymbol{n}$, with $\boldsymbol{n}$ is a unit vector showing the direction of the rotation axis. The equation of motion, in this case, has two supplementary terms [12]: the first one is "Centripetal acceleration", which may be written as $\boldsymbol{\Omega} \times (\boldsymbol{\Omega} \times \boldsymbol{u})$. This term arises due to time-varying motion only. The second term is "Corioli's acceleration" that is given as $2\boldsymbol{\Omega} \times \dot{\boldsymbol{u}}$. Accordingly, the principal equations for magneto-thermoelastic structure in the context of Lord-Shulman model may be written as (without heat sources, body forces and body couples):

(a) The equation of motion for (L-S) model with Lorentz force $\boldsymbol{f}$ and temperature T in rotating media may be written as:

$$\boldsymbol{\nabla} \cdot \boldsymbol{\sigma} + \boldsymbol{f} = \rho[\ddot{\boldsymbol{u}} + \{\boldsymbol{\Omega} \times (\boldsymbol{\Omega} \times \boldsymbol{u}\} + (2\boldsymbol{\Omega} \times \dot{\boldsymbol{u}})] - \gamma \boldsymbol{\nabla} \boldsymbol{T} \tag{1}$$

Where

$$\boldsymbol{f} = \boldsymbol{J} \times \boldsymbol{B} \tag{2}$$

(b) The simplified linear equations of Maxwell (in S.I. units) for electrodynamics of steadily dynamic solid for a homogeneous and electrically conducting structure are:

$$\boldsymbol{curl\,H} = \boldsymbol{J}, \quad \boldsymbol{curl\,E} = -\dot{\boldsymbol{B}}, \quad \boldsymbol{div\,B} = 0, \quad \nabla \cdot \boldsymbol{D} = \rho_e, \tag{3}$$

Considering the relationships between the vector of magnetic induction $\boldsymbol{B}$ and the vector of total magnetic field $\boldsymbol{H}$ is given as $\boldsymbol{B} = \mu_e \boldsymbol{H}$. So, in our case we will assume that $\rho_e = 0$ in Eqs. (3).

(c) The generalized heat conduction equation based on the (L-S) model takes the form:

$$\boldsymbol{K} \boldsymbol{\nabla}^2 \boldsymbol{T} = T_o \gamma (1 + t_o \frac{\partial}{\partial t}) \boldsymbol{\nabla} \dot{\boldsymbol{u}} + \rho C_e (\dot{\boldsymbol{T}} + t_o \ddot{\boldsymbol{T}}) \tag{4}$$

(d) The constitutive relation between the strain and the displacement are:

$$\varepsilon = \frac{1}{2} (\boldsymbol{\nabla u} + \boldsymbol{\nabla u}^T) \tag{5}$$

(e) Constitutive relation between stress, strain and thermal fields are:

$$\boldsymbol{\sigma} = \boldsymbol{c}\, \boldsymbol{\varepsilon} - \gamma (\boldsymbol{T} - T_o) \tag{6}$$

(f) The generalized Ohm's law is:

$$\boldsymbol{J} = \sigma_o [\boldsymbol{E} + (\dot{\boldsymbol{u}} \times \boldsymbol{B})] \tag{7}$$

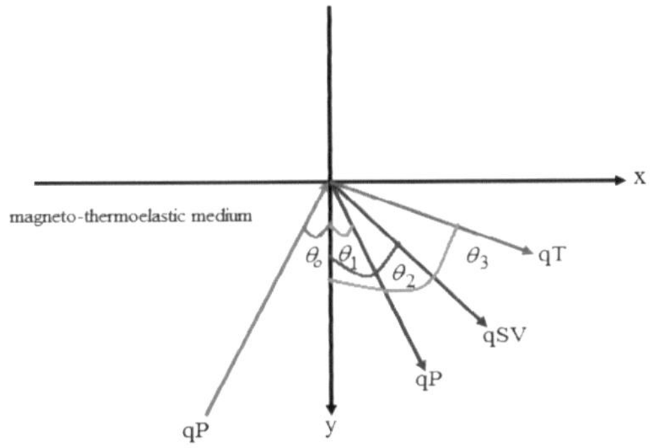

Figure 1. Geometry of the problem.

where the elastic part of the stress tensor σ is in equation (6), for the transversely isotropic materials, it may be seen as in [18]. The superimposed dot gives the derivatives of time and the symbols manifesting in the previous relations are listed in Table 1. The complete imagination of the problem is presented in Figure 1.
We consider the problem of a magneto-thermoelastic half-space ($y \leq 0$) with neglecting the conduction current $\boldsymbol{J}$ because of the small effect that is of temperature gradient.
We assume that $\boldsymbol{H} = \boldsymbol{h} + \boldsymbol{H}_o$ with $\boldsymbol{H} = (0, 0, H_o)$. The perturbed magnetic field $\boldsymbol{h}$ is so small such that the multiplication of $\boldsymbol{h}$, $\boldsymbol{u}$ and their derivatives may be ignored during linearizing of the equations. Also, the magnetic field for a constant intensity $\boldsymbol{H} = (0, 0, H_o)$ will be considered to act parallel to the bounding plane (take as the direction of the z-axis). Therefore, all the above-mentioned quantities will be functions of the time variable t and of the coordinates x and y [7].
The solution of equations (8), (9) and (10) can be chosen for a harmonic wave propagated for (L-S) model contains only one thermal relaxation time $t_o > 0$ in the following form:

$$\{u, v, T\} = (A, B, C)\exp\left[ik(x \sin\theta - y\cos\theta - ct)\right] \tag{8}$$

Where the wave normal locates in the xy-plane, and displays an angle θ with the y-axis.
From Eq. Therefore, we may have a system of homogeneous equations consisting of three Eqs. as:

$$A\left[D_1 - \zeta\Omega^*\right] + B\left[D_2 + 2\frac{\Omega}{\omega}i\zeta\right] + \tau C^* \sin\theta = 0, \tag{9}$$

$$A\left[D_2 - 2\frac{\Omega}{\omega}i\zeta\right] + B\left[D_3 + \zeta\Omega^*\right] - \tau_n\bar{\gamma}C^* \cos\theta = 0, \tag{10}$$

$$i\lambda(\varepsilon\zeta\sin\theta)A - i\lambda(\bar{\gamma}\varepsilon\zeta\cos\theta)B + \frac{1}{\omega}(\zeta - D_5)C^* = 0, \tag{11}$$

Where we have used

$$D_1 = (c_{11} + \mu_e H_o^2)\sin^2\theta + c_{66}\cos^2\theta, \quad D_2 = -(c_{12} + c_{66} + \mu_e H_o^2)\sin\theta\cos\theta,$$

$$D_3 = (c_{22} + \mu_e H_o^2)\cos^2\theta + c_{66}\sin^2\theta, \quad D_4 = K_1\sin^2\theta + K_2\cos^2\theta,$$

By using the following of dimensionless quantities

$$\zeta = \rho c^2, \omega = Kc, \quad \bar{\gamma} = \frac{\gamma_2}{\gamma_1}, \quad \Omega^* = 1 + \frac{\Omega^2}{\omega^2}, \quad C^* = \frac{\omega\gamma_1 C}{k}, \quad \varepsilon = \frac{T_o\gamma_1^2}{\rho C_e}, \quad \lambda_n = nt_o + \frac{i}{\omega},$$

$$\lambda_1 = t_o + \frac{i}{\omega}, \quad D_5 = \frac{D_4}{\tau_1 C_e}, \quad \alpha = \frac{(i/\omega)}{t_o + (i/\omega)} = \frac{(i/\omega)}{\lambda_1}.$$

To get the non-trivial solutions of the system of Eqs. (9), (10) and (11), the determination of the factor of matrix must vanish, therefore,

$$|L_{ij}|=0 \tag{12}$$

Where

$L_{11}=D_1-\zeta\Omega^*$, $L_{12}=D_2+2i\zeta\dfrac{\Omega}{\omega}$, $L_{13}=\lambda_n\sin\theta$, $L_{12}=D_2-2i\zeta\dfrac{\Omega}{\omega}$, $L_{22}=D_3-\zeta\Omega^*$,
$L_{23}=\lambda_n\bar{\gamma}\cos\theta$, $L_{31}=i\omega\alpha\varepsilon\zeta\sin\theta$, $L_{32}=-i\omega\alpha\bar{\gamma}\varepsilon\zeta\cos\theta$, $L_{33}=\zeta-D_5$.
This yields the cubic equation in ζ as:

$$A_0\zeta^3+A_1\zeta^2+A_2\zeta+A_3=0 \tag{13}$$

Where,

$A_0=4\left(\dfrac{\Omega}{\omega}\right)^2-\Omega^{*2}$, $A_1=\Omega^*(D_1+D_3)+(\Omega^{*2}-4(\dfrac{\Omega}{\omega})^2)D_5-i\lambda_n\varepsilon\omega\alpha(\sin^2\theta+\bar{\gamma}^2\cos^2\theta)\Omega^*$

$A_2=D_2^2-D_1D_3+\Omega^*(D_1D_5+D_3D_5)+i\lambda_n\varepsilon\omega\alpha(2\bar{\gamma}D_2\sin\theta\cos\theta+\bar{\gamma}^2D_1\cos^2\theta+D_3\sin^2\theta)$,

$A_3=(D_1D_3-D_2^2)D_5$,

Here, the three roots $\zeta_j=\rho c_j^2$, (j=1,2,3) of equation (13) corresponding to the phase velocities c_j, (j=1,2,3) of QP, QSV and QT waves, respectively, which is has a complex value. If we use the relation $c_j^{-1}=V_j^{-1}-i\omega q_j$, ($j$=1,2,3), then it will be obvious that V_j and q_j are the velocity of propagation and the coefficients of attenuation of the QP, QSV and QT waves [16].

3. Boundary conditions

The required boundary conditions at the free surface $y=0$ are vanishing of the normal stresses, tangential stresses and normal component of the heat flux vector, i.e.,

$$\sigma_{yy}+\sigma^*_{yy}=0,\ \sigma_{yx}+\sigma^*_{yx}=0 \quad \text{and} \quad T_{,y}=0 \text{ on } \quad y=0, \tag{14}$$

Where

$\sigma_{yy}=c_{12}u_{,x}+c_{22}v_{,y}-\beta_2(T+t_1\dot{T})$, $\sigma^*_{yy}=-\mu_eH_o^2(u_{,x}+v_{,y})$, $\sigma_{yx}=c_{44}(u_{,y}+v_{,x})$, $\sigma^*_{yx}=0$.

4. Reflection from free surface

In this section, we shall derive the relations between the reflection coefficients, when (QP or QT or QSV) wave becomes incident at a thermoelastic solid half-space $(y\geq 0)$ with thermally insulated and stress-free surface $y=0$. The positive $y-$axis is taken into the half-space. For the incident wave at the free surface, there will be three reflected waves, i.e., reflected QP, reflected QT and reflected QSV. Accordingly, if the wave normal to the incident wave (QP or QT or QSV) makes angle θ_o with the positive direction of the y-axis and those of reflected QP, QT and QSV waves make θ_1, θ_2 and θ_3 with the same direction. The complete geometry showing the incident and reflected waves is depicted in Figure 1. The appropriate displacement components and temperature field which must satisfy the boundary conditions (14) at y=0 are as below:

$$u=A_oe^{\eta_o}+A_1e^{\eta_1}+A_2e^{\eta_2}+A_3e^{\eta_3}, \tag{15}$$

$$v=F_oA_oe^{\eta_o}+F_1A_1e^{\eta_1}+F_2A_2e^{\eta_2}+F_3A_3e^{\eta_3}, \tag{16}$$

$$T=G_oA_oe^{\eta_o}+G_1A_1e^{\eta_1}+G_2A_2e^{\eta_2}+G_3A_3e^{\eta_3}, \tag{17}$$

With

$\eta_o=ik_o(x\sin\theta_o-y\cos\theta_o-c_ot)$, $\eta_l=ik_l(x\sin\theta_l+y\cos\theta_l-c_lt)$, $l=1,2,3$

Where c_o is the velocity of the incident QP or QT or QSV wave, k_l, (l=0,1,2,3) are complex wave numbers, $c_l = \zeta_l / \rho$ ($l = 0,1,2,3$).

Substituting from Eqs. (15-17) into Eq.(12) and Eq. (13), we get in the case of incident wave

$$F_o = -N_1 / N_2, \qquad G_o = (k_o / \omega\lambda_n)(N_3 / N_2) \tag{18}$$

Where

$$N_1 = \gamma_1 \sin\theta_o[(D_{2o} - 2i(\Omega/\omega)\zeta_o] + \gamma_2 \cos\theta_o(D_{1o} - \zeta_o\Omega^*),$$
$$N_2 = \gamma_2 \cos\theta_o[(D_{2o} + 2i(\Omega/\omega)\zeta_o] + \gamma_1 \sin\theta_o(D_{3o} - \zeta_o\Omega^*),$$
$$N_3 = [(4(\Omega/\omega)^2 - \Omega^{*2})\zeta_o^2 + \Omega^*(D_{1o} + D_{3o})\zeta_o + D_{2o}^2 - D_{1o}D_{3o}],$$

With

$$D_{1o} = (c_{11} + \mu_e H_o^2)\sin^2\theta_o + c_{66}\cos^2\theta_o, \qquad D_{2o} = -(c_{12} + c_{66} + \mu_e H_o^2)\sin\theta_o\cos\theta_o,$$
$$D_{3o} = (c_{22} + \mu_e H_o^2)\cos^2\theta_o + c_{66}\sin^2\theta_o.$$

In the case of reflected waves, one may get the following displacement components and temperatures field which are suitable to satisfy the boundary conditions (14) at y=0.

$$u = A_\ell e^{\eta_\ell}, \; v = F_\ell A_\ell e^{\eta_\ell} \text{ and } T = G_\ell A_\ell e^{\eta_\ell}, \; \ell = 1,2,3 \tag{19}$$

With

$$F_\ell = -N_{1\ell} / N_{2\ell}, \quad G_\ell = -(k_\ell / \omega\lambda_n)(N_{3\ell} / N_{2\ell}), \tag{20}$$

$$N_{1\ell} = \gamma_1 \sin\theta_\ell[(D_{2\ell} - 2i(\Omega/\omega)\zeta_\ell] - \gamma_2 \cos\theta_\ell(D_{1\ell} - \zeta_\ell\Omega^*),$$
$$N_{2\ell} = \gamma_2 \cos\theta_\ell[(D_{2\ell} + 2i(\Omega/\omega)\zeta_\ell] - \gamma_1 \sin\theta_\ell(D_{3\ell} - \zeta_\ell\Omega^*),$$
$$N_{3\ell} = [(4(\Omega/\omega)^2 - \Omega^{*2})\zeta_\ell^2 + \Omega^*(D_{1\ell} + D_{3\ell})\zeta_\ell + D_{2\ell}^2 - D_{1\ell}D_{3\ell}],$$

Where

$$D_{1\ell} = (c_{11} + \mu_e H_o^2)\sin^2\theta_\ell + c_{66}\cos^2\theta_\ell, \qquad D_{2\ell} = (c_{12} + c_{66} + \mu_e H_o^2)\sin\theta_\ell\cos\theta_\ell,$$

$$D_{3\ell} = (c_{22} + \mu_e H_o^2)\cos^2\theta_\ell + c_{66}\sin^2\theta_\ell.$$

Using Eqs. (15-17) which corresponding to incident and reflected waves in the boundary conditions (14) yield

$$\begin{aligned} &-(c_{22} - \mu_e H_o^2)(F_o A_o(ik_o\cos\theta_o) - F_1 A_1(ik_1\cos\theta_1) - F_2 A_2(ik_2\cos\theta_2) \\ &- F_3 A_3(ik_3\cos\theta_3)) + (c_{12} - \mu_e H_o^2)(A_o(ik_o\sin\theta_o) + A_1(ik_1\sin\theta_1) + \\ &A_2(ik_2\sin\theta_2) + A_3(ik_3\sin\theta_3)) - \gamma_2(1 - it_1\omega)(G_o A_o + G_1 A_1 + G_2 A_2 + G_3 A_3) = 0, \end{aligned} \tag{21}$$

$$\begin{aligned} &-ik_o(A_o\cos\theta_o - A_o F_o\sin\theta_o) + ik_1(A_1\cos\theta_1 + A_1 F_1\sin\theta_1) \\ &+ ik_2(A_2\cos\theta_2 + A_2 F_2\sin\theta_2) + ik_3(A_3\cos\theta_3 + A_3 F_3\sin\theta_3) = 0, \end{aligned} \tag{22}$$

$$-ik_o(G_o A_o\cos\theta_o) + ik_1(G_1 A_1\cos\theta_1) + ik_2(G_2 A_2\cos\theta_2) + ik_3(G_3 A_3\cos\theta_3) = 0. \tag{23}$$

Since the phases of the waves should be the same for each values of x, then we obtain the following relations which must be valid

$\eta_o = \eta_1 = \eta_2 = \eta_3$,

and so

$k_o\sin\theta_o = k_1\sin\theta_1 = k_2\sin\theta_2 = k_3\sin\theta_3$

And
$k_o c_o = k_1 c_1 = k_2 c_2 = k_3 c_3$.
Therefore, the Snell's law may be deduced as

$$\frac{\sin\theta_o}{Vo} = \frac{\sin\theta_1}{V_1} = \frac{\sin\theta_2}{V_2} = \frac{\sin\theta_3}{V_3}, \tag{24}$$

Multiplying Eqs. (24-26) by the factor $\frac{i}{k_o A_o}$, using Snell's law (27). putting

$$\frac{A_i}{A_o} = W_i, \tag{25}$$

one may get the following system of three non-homogenous equations,

$$\sum_{j=1}^{3} a_{ij} W_j = b_i, \quad i = 1,2,3 \tag{26}$$

Where

$$a_{1L} = F_L(c_{22} - \mu_e H_o^2)[(\frac{V_o}{V_L})^2 - \sin^2\theta_o]^{1/2} + (c_{12} - \mu_e H_o^2)\sin\theta_o + \frac{\omega\lambda_n\gamma_2}{k_o} G_L,$$

$$a_{2L} = [(\frac{V_o}{V_L})^2 - \sin^2\theta_o]^{1/2} + F_L \sin\theta_o, \qquad a_{3L} = G_L[(\frac{V_o}{V_L})^2 - \sin^2\theta_o]^{1/2},$$

$$b_1 = (c_{22} - \mu_e H_o^2) F_o \cos\theta_o - (c_{12} - \mu_e H_o^2)\sin\theta_o - \frac{\omega\lambda_n\gamma_2}{k_o} G_o,$$

$$b_2 = \cos\theta_o - F_o \sin\theta_o, \qquad b_3 = G_o \cos\theta_o, \quad L = 1,2,3.$$

From Eqs. (26). the analytical expressions of amplitude ratios W_1, W_2 and W_3 for incident QP waves may be obtained. While the analytical expressions of amplitude ratios Y_1, Y_2 and Y_3 for incident QSV are related to amplitude ratios W_j with the relations:

$$Y_1 = F_1^* W_1, \quad Y_2 = F_2^* W_2, \quad Y_3 = F_3^* W_3 \tag{27}$$

Furthermore, the analytical expressions of amplitude ratios Z_1, Z_2 and Z_3 for incident QT are related to amplitude ratios W_j with the relations:

$$Z_1 = G_1^* W_1, \quad Z_2 = G_2^* W_2, \quad Z_2 = G_2^* W_2 \tag{28}$$

Where

$$F_j^* = \frac{F_j}{F_o}, \quad G_j^* = \frac{G_j}{G_o}, \quad j=1,2,3. \tag{29}$$

It is noted that the definitions of the previous amplitude ratios are controversial. The reason is that only the amplitude ratios of the horizontal parts of the mechanical displacement have been taken into account and no contribution comes from the vertical displacement nor from the temperature fields. A general expression of the amplitude ratios, in which all the contributions coming from both the vertical displacement and the temperature fields will be considered in the future works (see for example [20]).

5. Numerical results and discussion

For numerical calculations, the material of Zinc (Zn) is chosen which has the following physical information [22].

$c_{11} = 1.628\times10^{11}\,Nm^{-2}$ $c_{12} = 0.362\times10^{11}\,Nm^{-2}$

$c_{22} = 1.628\times10^{11}\,Nm^{-2}$ $C_e = 3.9\times10^{2}\,JKg^{-2}\deg^{-1}$

$\gamma_1 = 5.75\times10^{6}\,Nm^{-2}\deg^{-1}$ $\gamma_2 = 5.07\times10^{6}\,Nm^{-2}\deg^{-1}$

$K_1 = 1.24\times10^{2}\,Wm^{-2}\deg^{-1}$ $K_2 = 1.24\times10^{2}\,Wm^{-2}\deg^{-1}$

$\rho = 7.14\times10^{3}\,Kgm^{-3}$ $T_o = 296$ K

The absolute values of the reflection coefficients for various quasi-plane waves are numerically considered against the incidence angle of (QP or QSV or QT) waves for various values of the thermal relaxation time and rotation variables from Eqs. (26), (27), (28) and (29). These changes of the reflection coefficients are given graphically in the Figures 2 and 3. Therefore, with the help of the Figures, one may summarize the following observations:

(i) From Figure 2a, it is noted that W_1 starts with the value 1.049 when $\theta_0 = 0^\circ$ then decrease to reach the minimum value 0.78 at $\theta_0 = 28^\circ$ wherever there is no effect for the variation to $t_\circ$. After that, W_1 grows monotonically from 1.049 at $\theta_o = 29^\circ$ up to the maximum value 1.724 at $\theta_o = 66^\circ$ and then reduces down to the minimum value 1.413 at $\theta_o = 90^\circ$. The influence of the $t_\circ$ appears in the period $\theta_o = (45^\circ - 90^\circ)$, where there is an inverse proportion to the increased of thermal relaxation time $t_\circ$.

Figures. 2b and 2c, display the reflection coefficients W_2 and W_3 as a function of θ_o. In these situations, the specific attitudes of these Figures are roughly analogous to the standard normal distribution in probability. It is noticed that the maximum of the modulus of W_1 is 0.268 that occurs around $\theta_o = 54^\circ$, and similarly, the curves corresponding to the change of $t_\circ$ offer relevant variations only round the maximum. The same effect also happens for W_3, but the effect of the thermal relaxation time be less and it becomes negligible when $\theta_o = 90^\circ$.

(ii) For the second group, the results are given in Figures. (3a, 3b, 3c), where changes of the dimensionless rotation come as $\Omega' = 4.0, 4.4, 4.8$, with the fixed values of both $H_o = 10^6$ A/m and $t_\circ = 3$ PS. In this case, the Figures 3a displays the effectiveness of variation in the rotation Ω' (respectively in, 3b and 3c), on the reflection coefficient W_1 as function of θ_o. It has been observed that W_1 has the same behavior as in Figure 2a. In details, the value of W_1 decreases in the two intervals $\theta_o = (0^\circ, 29^\circ)$ and $\theta_o = (61^\circ, 90^\circ)$. However, the opposite happens in the interval $\theta_o = (30^\circ, 60^\circ)$. An absolute maximum for W_1 becomes visible about $\theta_o = 62^\circ$. Figures. 3b and 3c have a similar attitude as Figures. 2b and 2c, respectively. However, reflection coefficients are directly proportional to an increased rotation.

6. Conclusion

This mathematical model illustrates the characteristics of reflection of quasi-plane wave propagation in a transversely isotropic magneto-thermoelastic structure. The solution is given for suitable boundary conditions in the context of the generalized theory of thermo elasticity based on Lord-Shulman Model. In this work existence of three quasi-plane waves QP, QSV and QT waves are found. Furthermore, the reflection coefficients ratios depend upon the rotation, angle of reflection, angle of incidence and anisotropy in the structure. This research is significant to solve problems with parameters like anisotropic elastic structure and rotation coexist, for example, the seismic wave propagation inside the earth, the characteristics of earthquakes, nuclear arms and nuclear devices and underground nuclear explosions. Furthermore, the methods which are used in this paper may be applied in some recent interesting practical works and future research directions (see for example [23] -[25]).

7. References

[1] Muller I M 1971, The coldness. A universal function in thermoelastic bodies. Arch. Ration. Mech. Anal, 41, 319–327.

[2] Green A E and Lindsay K A 1972, Thermoelasticity, J. Elast., 2 1–7.

[3] Suhubi E S 1975, Thermoelastic solids. In: Eringen, A.C. (eds) Continuum Physics, Academic Press, New York.

[4] Lord H and Shulman Y 1967, A generalized dynamical theory of thermoelasticity. J. Mech. Phys. Solid 15, 299–309.

[5] Dhaliwal R S and Sherief H H 1980, Generalized thermoelasticity for anisotropic media, Quart. Appl. Math., 38, 1-8.

[6] Hetnarski R B and Ignaczak J 1999, Generalized thermoelasticity, *J. Therm. Stresses* **22**, 451–476.

[7] Paria G 1967, Magneto-elasticity and magneto-thermoelasticity, Adv. Appl. Mech. **10**, 73–112.

[8] Abd-alla A N 1999, Nonlinear constitutive equations for thermo-electroelastic materials. *Mechanics Research Communications* **24,** 335–346.

[9]] Abd-alla A N, and Abo-Dahab S M 2008, The influence of the viscosity and the magnetic field on reflection and transmission of waves at interface between magneto-viscoelastic materials, *Meccanica* **43,** 437-448.

[10] Abd-alla A N, Yahia A A and Abo-Dahab S M 2003, On Reflection of the Generalized Magneto-thermo-viscoelastic Plane Waves, *Chaos, Solitons and Fractals* **16,** 211–231.

[11] Abd-alla A N 2000, Relaxation effects on reflection of generalized magneto-thermo-elastic waves, *Mechanics Research Communications* **27,** 591-600.

[12] Othman M I A and Song Y Q 2006, The effect of rotation on the reflection of magneto-thermo-elastic waves under thermoelasticity without energy dissipation, *Acta Mech.* **184,** 189–204.

[13] Othman M I A and Song Y Q 2008, Reflection of magneto-thermoelastic waves from a rotating elastic half-space, Int. J. Engng. Sci. 46, 459-474.

[14] Ting T C T 2004, Surface waves in a rotating anisotropic elastic half-space, Wave Motion 40, 329–346.

[15] Kumar R and Singh M 2009. Effects of rotating and imperfection on reflection and transmission of plane waves in anisotropic generalized thermoelastic media, *J. Sound Vibr.* **324**, 773–797.

[16] Singh B and Yadav A K 2012, Reflection of plane waves in a rotating transversely isotropic magneto-thermoelastic solid half space, *J. of Theor. Appl. Mech., Sofia*, **42,** 33-60.

[17] Abd-alla A N, Giorgio I, Galantucci L, Hamdan A M and del Vescovo D 2016, Wave reflection at a free interface in an anisotropic pyroelectric medium with nonclassical thermoelasticity, *Continuum Mechanics and thermodynamics*, 28 (1-2), 67-84.

[18] Abd-alla A N, Hamdan A M, Giorgio I and Del Vescovo D 2014, The mathematical model of reflection and refraction of longitudinal waves in thermo-piezoelectric materials, *Archive of Applied Mechanics* **84**, 1229-1248.

[19] Abd-alla A N, Alshaikh F A and Al-Hossain A Y 2012, The reflection phenomena of quasi-vertical transverse waves in piezoelectric medium under initial stresses, *Meccanica* **47,** 731-744.

[20] Abd-alla A N and Alshaikh F A 2009, Reflection and refraction of plane quasi-longitudinal waves at an interface of two piezoelectric media under initial stresses, *Archive of Applied Mechanics* **79,** 843-857.

[21] Abbas A I, Abd-alla A N and Othman M I A 2011, Generalized Magneto-thermoelasticity in a Fiber-Reinforced Anisotropic Half-Space *International Journal of Thermophysics*, **32**, 1071-1085

[22] Sharma J, Pathania V 2004, Generalized thermoelastic waves in anisotropic plates sandwiched between liquid layers, *J. Sound and Vibration*, **278**, 383–411.

[23] Abbas I A, Abd-alla A N and Alshaikh F 2017, Analytical and computational solution of three-dimensional thermoelastic interactions in porous material with temperature-dependent properties, *Journal of Computational and Theoretical Nanoscience* **14(12)** 5952-5957.

[24] Abd-alla A N, Hamdan A M, Almarashi A A and Battista A (2017), The mathematical modeling for bulk acoustic wave propagation velocities in transversely isotropic piezoelectric materials, *Mathematics and Mechanics of Solids*, **22(4)** 823–836.
[25] Abd-alla A N, Othman M I A and Hasbullah N F (2017), Shear horizontal waves in composite materials: Behavior under rotation and initial stress, Mathematics and Mechanics of Solids, 1–13. First Published October 6, https://doi.org/10.1177/1081286517730682

Acknowledgment
The author is grateful to the Deanship of Scientific Research of Jazan University for supporting financially this work under the Waed Grant No. 4/38.

Table 1. Nomenclature

ρ	mass density
ρ_e	charge density
$\boldsymbol{u}$	mechanical displacement
$\boldsymbol{\varepsilon}$	strain tensor
$\boldsymbol{\sigma}$	stress tensor
c	fourth order tensor tiffnesss
$\boldsymbol{B}$	magnetic induction vector
$\boldsymbol{H}$	total magnetic field vector
E	electric field vector
$\boldsymbol{f}$	electromagnetic body force (Lorentz force)
$\boldsymbol{J}$	electric current density vector
$\boldsymbol{H}_o$	perturbed magnetic field vector $\boldsymbol{h}$
T	temperature
T_0	reference uniform temperature of the body
K	heat conduction second order tensor
$\boldsymbol{\Omega}$	rotation vector
$\boldsymbol{\gamma}$	thermal elastic coupling second order tensor
C_e	specific heat at constant strain
$t_\circ$	thermal relaxation time
n	vector of the symmetric axis
N	second order tensor $\mathbf{n} \otimes \mathbf{n}$
I	identity tensor of second order
$\lambda, \mu, \mu_0, \alpha, \beta$	elastic constitutive parameters
μ_e	magnetic permeability of the medium
σ_o	electric conductivity of the medium
ω	circular frequency
c	complex phase velocity
θ	angle of propagation measured from the normal to the half-space

IOP Conf. Series: Materials Science and Engineering **348** (2018) 012018 doi:10.1088/1757-899X/348/1/012018

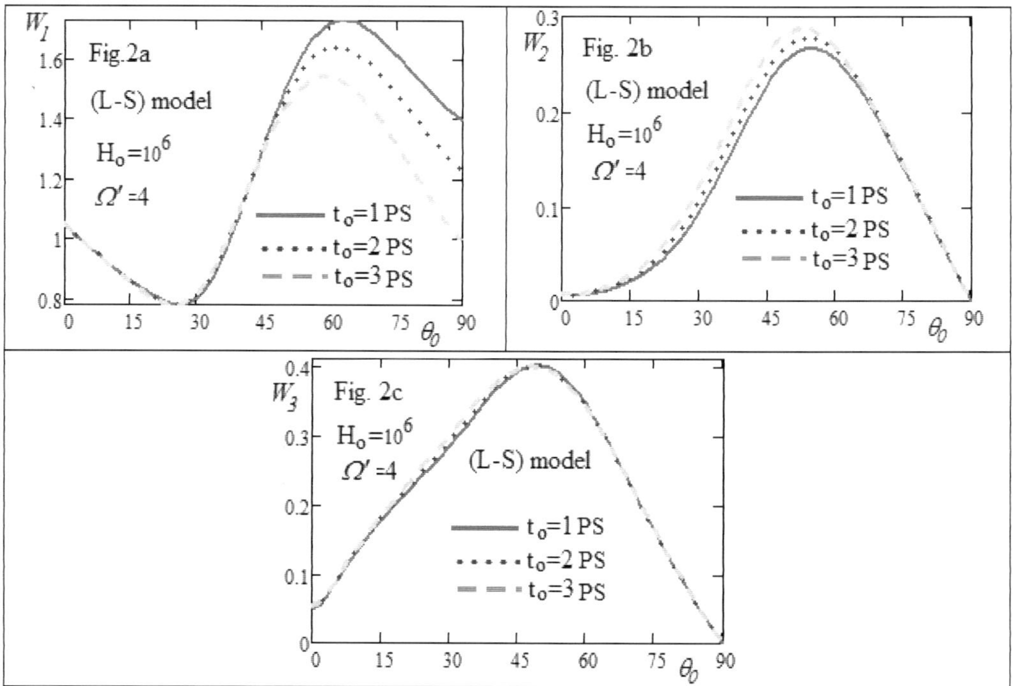

Figures (2a,2b,2c). Represent the reflection coefficients W_1 (2), W_2 (3), and W_3 (4) against θ_0 for different values of the relaxation time $t_\circ$.

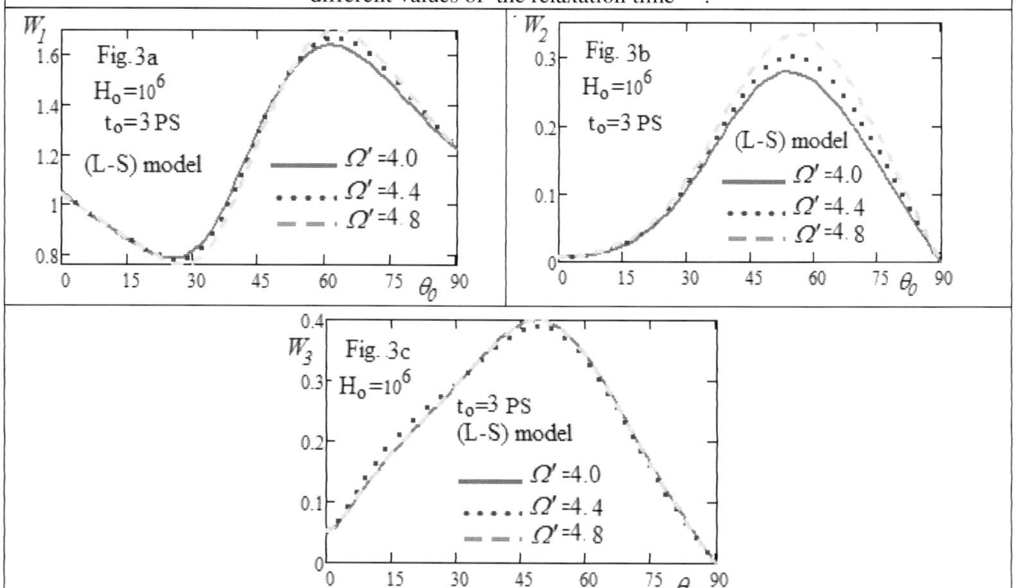

Figures (3a,3b,3c). Represent the reflection coefficients W_1 (5), W_2 (6), and W_3 (7) against θ_0 for different values of the rotation Ω'.

Generalized thermoelastic interaction in an isotropic solid cylinder without energy dissipation

Fatimah Alshaikh

Department of Mathematics, Faculty of Science, Jazan University, Jazan 2097, Saudi Arabia.
E-mail: fatimah@jazanu.edu.sa

Abstract. In this paper, we constructed the generalized thermoelastic equations of an isotropic solid cylinder. The formulation is applied in the context of Green and Naghdi theory of types II (without energy dissipation). The material of the cylinder is supposed to be homogeneous isotropic both mechanically and thermally. The governing equations have been written in the form of a vector-matrix differential equation in the Laplace transform domain, which is then solved by an eigenvalue approach. Numerical results for the temperature distribution, displacement and radial stress are represented graphically.

1. Introduction

The classical theory of thermoelasticity was studied by Carlson [1] in terms of generalizations and modifications into various thermoelastic models that can run under label hyperbolic thermoelasticity, see the survey of Chandrasekharaiah [2] and Hitnarski and Ignazack [3]. The notation hyperbolic reflects the fact that thermal waves are modelled, avoiding the physical paradox of infinite propagation speed of the classical models. Lately, Green and Naghdi (GN) [4-5] reinvestigated the principal hypotheses of thermo-mechanics that established a distinct entropy equality rather than the familiar entropy inequality. Model II of (GN) theory anticipates a limited speed for the propagation of heat and includes no energy dissipation which is known as thermoelasticity without energy dissipation.

During the second half of twentieth century, non-isothermal problems of the theory of elasticity become increasingly important. This is due to their many applications in widely diverse fields. First, the high velocities of modern aircraft give rise to aerodynamic heating, which produces intense thermal stresses that reduces the strength of the aircraft structures. Second, in the nuclear field, the extremely high temperature and temperature gradients originating inside nuclear reactors influence their design and operations.

The counterparts of our problem in the contexts of the uncoupled thermoelasticity theory, the coupled thermoelasticity theory, the Green-Lindsay theory (GL-theory) [6] and the Lord-Shulman theory (LS-theory) [7] have been considered by Othman et al. [8-10]. Furthermore, related generalized Magneto-thermoelasticity problems in different hypotheses have been studied by Abbas et al. [11-13]. The effective method of eigenvalue approach is used to get the solutions of variously coupled thermoelasticity in an isotropic or an anisotropic media associated with additional fields that has been studied in [14-18].

In the present paper, we consider a problem of generalized thermoelasticity of an isotropic solid cylinder based on Green-Naghdi theory of type II. The non-dimensional equations are handled by employing an analytical-numerical techniques based on Laplace transform and eigenvalues approach.

IOP Conf. Series: Materials Science and Engineering **348** (2018) 012019 doi:10.1088/1757-899X/348/1/012019

Numerical results for the temperature distribution, displacement and radial stress are shown graphically.

2. Basic equations

Following Green, and Naghdi [4, 5], the system of equations that include the displacement, the stress, the strain and the temperature for a linear, homogeneous and isotropic thermoelastic continuum take the following form

The equation of heat conduction

$$K^{*}T_{,ii} + K\dot{T}_{,ii} = \rho C_E \ddot{T} + \gamma T_o \ddot{u}_{i,i}, \tag{1}$$

The equations of motion

$$(\lambda+\mu)u_{j,ij} + \mu u_{i,jj} - \gamma T_{,i} = \rho\ddot{u}_i, \tag{2}$$

The constitutive equations are given by

$$\tau_{ij} = \lambda u_{i,i}\delta_{ij} + \mu\left(u_{i,j} + u_{j,i}\right) - \gamma T\delta_{ij}. \tag{3}$$

where C_E is the specific heat at constant strain, λ, μ are Lame's constants, $\gamma=(3\lambda+2\mu)\alpha_t$, α_t is the coefficient of linear thermal expansion, T is the temperature above reference temperature T_0, K^* are material constant characteristic of the theory. In a cylindrical coordinate system (r,θ,z), for the axially symmetric problem, $u_r=u_r(r,z,t)$, $u_\theta=0$, $u_z=u_z(r,z,t)$. Furthermore, if only axisymmetric plane strain problem is considered, we have $u_r= u(r,t)$ and $u_\theta=u_z=0$.

The strain-displacement relations are

$$e_{rr} = \frac{\partial u}{\partial r},\ e_{\theta\theta} = \frac{u}{r},\ e_{zz} = e_{rz} = e_{r\theta} = e_{\theta z} = 0. \tag{4}$$

The stress-strain relations are

$$\tau_{rr} = 2\mu\frac{\partial u}{\partial r} + \lambda\left(\frac{\partial u}{\partial r} + \frac{u}{r}\right) - \gamma T, \tag{5}$$

$$\tau_{\theta\theta} = 2\mu\frac{u}{r} + \lambda\left(\frac{\partial u}{\partial r} + \frac{u}{r}\right) - \gamma T, \tag{6}$$

It is assumed that there are no body forces and heat sources in the medium, the equation of motion and energy equation has the form:

$$\frac{\partial\tau_{rr}}{\partial r} + \frac{\tau_{rr} - \tau_{\theta\theta}}{r} = \rho\frac{\partial^2 u}{\partial t^2}, \tag{7}$$

$$K^{*}\frac{1}{r}\frac{\partial}{\partial r}\left(r\frac{\partial T}{\partial r}\right) = \rho C_E\frac{\partial^2 T}{\partial t^2} + \gamma T_o\frac{\partial^2}{\partial t^2}\left(\frac{\partial u}{\partial r} + \frac{u}{r}\right). \tag{8}$$

It is convenient to change the preceding equations into the dimensionless forms. To do this, the dimensionless parameters are introduced as

$$(r^{o}, u^{o}) = \frac{(r,u)}{c_1\omega_1},\quad t^{o} = \frac{t}{\omega_1},\quad \left(\tau^{o}_{rr}, \tau^{o}_{\theta\theta}\right) = \frac{1}{\mu}\left(\tau_{rr}, \tau_{\theta\theta}\right),\quad \theta^{o} = \frac{\gamma T}{\rho c_1^2}, \tag{9}$$

where, $c_1^2 = \dfrac{\lambda + 2\mu}{\rho}$, $\omega_1 = \dfrac{K}{\rho C_E c_1^2}$.

From (9) into equations (5)-(8), it is easy to obtain (after dropping the superscript $^\circ$ for convenience) the following:

$$\frac{\partial^2 u}{\partial r^2} + \frac{1}{r}\frac{\partial u}{\partial r} - \frac{u}{r^2} - \frac{\partial \theta}{\partial r} = \frac{\partial^2 u}{\partial t^2}, \tag{10}$$

$$\varepsilon_2 \frac{1}{r}\frac{\partial}{\partial r}\left(r\frac{\partial T}{\partial r}\right) = \frac{\partial^2 T}{\partial t^2} + \varepsilon_1 \frac{\partial^2}{\partial t^2}\left(\frac{\partial u}{\partial r} + \frac{u}{r}\right). \tag{11}$$

$$\tau_{rr} = (2+\beta)\frac{\partial u}{\partial r} + \beta\frac{u}{r} - (2+\beta)\theta, \tag{12}$$

$$\tau_{\theta\theta} = \beta\frac{\partial u}{\partial r} + (2+\beta)\frac{u}{r} - (2+\beta)\theta, \tag{13}$$

$$\beta = \frac{\lambda}{\mu},\ \varepsilon_1 = \frac{\gamma^2 T_0}{\rho^2 c_1^2 C_E},\ \varepsilon_2 = \frac{K^*}{\rho c_1^2 C_E}.$$

From preceding description, the initial and boundary conditions may be expressed as

$$u(r,0) = \frac{\partial u(r,0)}{\partial t} = 0, \theta(r,0) = \frac{\partial \theta(r,0)}{\partial t} = 0, \tag{14}$$

$$-K\frac{\partial \theta(r,t)}{\partial r} = q_o \frac{t^2 e^{-\frac{t}{t_p}}}{16 t_p^2}, \sigma_{rr}(r,t) = 0, \tag{15}$$

where b is an outer radius of the hollow cylinder, q_o is a constant, t_p is a characteristic time of the pulse heat flux.

3. Governing equations in the laplace transform domain

Applying the Laplace transform for equations (10)-(15) defined by the formula

$$\bar{f}(s) = L[f(t)] = \int_0^\infty f(t)e^{-st}dt. \tag{16}$$

Hence, we obtain the following system of differential equations

$$\frac{d^2\bar{u}}{dr^2} + \frac{1}{r}\frac{d\bar{u}}{dr} - \frac{\bar{u}}{r^2} - \frac{d\bar{\theta}}{dr} = s^2\bar{u}, \tag{17}$$

$$(\varepsilon_2 + s)\left(\frac{d^2\bar{\theta}}{dr^2} + \frac{1}{r}\frac{d\bar{\theta}}{dr}\right) - s^2\bar{\theta} = \varepsilon_1 s^2\left(\frac{d\bar{u}}{dr} + \frac{\bar{u}}{r}\right), \tag{18}$$

$$\tau_{rr} = (2+\beta)\frac{d\bar{u}}{dr} + \beta\frac{\bar{u}}{r} - (2+\beta)\bar{\theta}, \tag{19}$$

$$\tau_{\theta\theta} = \beta\frac{d\bar{u}}{dr} + (2+\beta)\frac{\bar{u}}{r} - (2+\beta)\bar{\theta}, \tag{20}$$

$$\frac{\partial\bar{\theta}}{\partial r} = \frac{-q_0 t_p}{8(st_p+1)^3}, \bar{\sigma}_{rr} = 0, \tag{21}$$

Differentiating equation (18) with respect to r and using equation (17), it is easy and interesting to obtain

$$\frac{d^2}{dr^2}\left(\frac{d\bar{\theta}}{\partial r}\right) + \frac{1}{r}\frac{d}{dr}\left(\frac{d\bar{\theta}}{\partial r}\right) - \frac{1}{r^2}\left(\frac{d\bar{\theta}}{\partial r}\right) = \frac{\varepsilon_1 s^4}{\varepsilon_2} + \frac{s^2(1+\varepsilon_1)}{\varepsilon_2}\left(\frac{d\bar{\theta}}{\partial r}\right). \tag{22}$$

Equations (17) and (22) can be written in a vector-matrix differential equation as follows

$$L\vec{V} = A\vec{V}, \tag{23}$$

where $L \equiv \frac{d^2}{dr^2} + \frac{1}{r}\frac{d}{dr} - \frac{1}{r^2}$, $\vec{V} = \begin{bmatrix} \bar{u} & \frac{d\bar{\theta}}{dr} \end{bmatrix}^T$ and $A = \begin{bmatrix} m_{11} & m_{12} \\ m_{21} & m_{22} \end{bmatrix}$, and

$m_{11} = s^2$, $m_{12} = 1$, $m_{21} = \frac{s^4\varepsilon_1}{\varepsilon_2}$, $m_{22} = \frac{s^2(1+\varepsilon_1)}{\varepsilon_2}$.

4. Solution of the vector-matrix differential equation

Let us now proceed to solve equation (23) by the eigenvalue approach proposed by [29].
The characteristic equation of the matrix A takes the form

$$m_{11}m_{22} - m_{12}m_{21} - (m_{22} + m_{11})\lambda^2 + \lambda^4 = 0. \tag{24}$$

The roots of the characteristic equation (24) which are also the eigenvalues of matrix A are of the form $\lambda^2 = \lambda_1^2, \lambda^2 = \lambda_2^2$. The eigenvector $\vec{X} = [x_1, x_2]^T$, corresponding to eigenvalue λ^2 can be calculated as:

$$x_1 = m_{12},\ x_2 = \lambda^2 - m_{11}. \tag{25}$$

From equations (24), we can easily calculate the eigenvector $\vec{X}_j$, corresponding to eigenvalue $\lambda_j^2, j = 1, 2$. For further reference, we shall use the following notations:

$$\vec{X}_1 = [\vec{X}]_{\lambda^2=\lambda_1^2},\ \vec{X}_2 = [\vec{X}]_{\lambda^2=\lambda_2^2}. \tag{26}$$

The general solutions of equation (23) must be finite when the radius $r \to 0$ and this leads to:

$$\vec{V} = \vec{X}_1 B_1 I_1(\lambda_1 r) + \vec{X}_2 B_2 I_1(\lambda_2 r), \tag{27}$$

where B_1 and B_2 are constants to be determined from the boundary condition of the problem. Thus, the field variables can be written for r and s as:

$$\bar{u}(r,s) = x_1^1 B_1 I_1(\lambda_1 r) + x_2^1 B_2 I_1(\lambda_2 r), \tag{28}$$

$$\bar{\theta}(r,s) = -\frac{x_1^2}{\lambda_1} B_1 I_0(\lambda_1 r) - \frac{x_2^2}{\lambda_2} B_2 I_0(\lambda_2 r), \tag{29}$$

$$\tau_{rr} = B_1\left[(2+\beta)\left(\lambda_1 x_1^1 + \frac{x_2^1}{\lambda_1}\right) I_0(\lambda_1 r) - \frac{2x_1^1}{r} I_1(\lambda_1 r)\right] + B_2\left[(2+\beta)\left(\lambda_2 x_1^2 + \frac{x_2^2}{\lambda_2}\right) I_0(\lambda_2 r) - \frac{2x_1^2}{r} I_1(\lambda_2 r)\right], \tag{30}$$

$$\tau_{\theta\theta} = B_1\left[\left(\beta\lambda_1 x_1^1 + (\beta+2)\frac{x_2^1}{\lambda_1}\right) I_0(\lambda_1 r) + \frac{2x_1^1}{r} I_1(\lambda_1 r)\right] + B_2\left[\left(\beta\lambda_2 x_2^1 + (\beta+2)\frac{x_2^2}{\lambda_2}\right) I_0(\lambda_2 r) + \frac{2x_2^1}{r} I_1(\lambda_2 r)\right] \tag{31}$$

To complete the solution, constants B_1 and B_2 are required by using the boundary conditions (21) can be obtain

$$\begin{pmatrix} B_1 \\ B_2 \end{pmatrix} = \begin{pmatrix} R_{11} & R_{12} \\ R_{21} & R_{22} \end{pmatrix}^{-1} \begin{pmatrix} 0 \\ \frac{1}{s} \end{pmatrix}, \tag{32}$$

where

$$R_{11} = (2+\beta)\left(\lambda_1 x_1^1 + \frac{x_2^1}{\lambda_1}\right) I_0(\lambda_1 b) - \frac{2x_1^1}{a} I_1(\lambda_1 b),$$

$$R_{12} = (2+\beta)\left(\lambda_2 x_1^2 + \frac{x_2^2}{\lambda_2}\right) I_0(\lambda_2 b) - \frac{2x_1^2}{a} I_1(\lambda_2 b),$$

$$R_{21} = -\frac{x_2^1}{\lambda_1} I_0(\lambda_1 b),$$

$$R_{22} = -\frac{x_2^2}{\lambda_2} I_0(\lambda_2 b),$$

5. Numerical inversion of the Laplace transforms

For the final solution of displacement, temperature and stress distributions, a numerical inversion method was adopted based on the Riemann-sum approximation method which is used to obtain the numerical results. In this method, a function in the Laplace domain can be transformed into the time domain as:

$$\mathrm{H}(x,t) = \frac{e^{mt}}{t}\left(\frac{1}{2} Re[\bar{\mathrm{H}}(x,m)] + Re \sum_{q=0}^{N^*} (-1)^q \bar{\mathrm{H}}\left(x, m + \frac{iq\pi}{t}\right)\right), \tag{33}$$

where Re is the real part and i is the imaginary number unit. For faster convergence, numerical experiments stated that $m = \frac{4.7}{t}$ which satisfies the above relation.

6. Numerical results and discussion

The copper material was chosen for purpose of numerical evaluations and the constants of the problem were taken as follows [12]:

$\lambda = 7.76\times10^{10}(\mathrm{kg})(\mathrm{m})^{-1}(\mathrm{s})^{-2}, \mu = 3.86\times10^{10}(\mathrm{kg})(\mathrm{m})^{-1}(\mathrm{s})^{-2},$

$\mathrm{K} = 3.68\times10^{2}(\mathrm{kg})(\mathrm{m})(\mathrm{K})^{-1}(\mathrm{s})^{-3}, c_e = 3.831\times10^{2}(\mathrm{m})^{2}(\mathrm{K})^{-1}(\mathrm{s})^{-2},$

$\rho = 8.954\mathrm{x}10^{3}(\mathrm{kg})(\mathrm{m})^{-3}, T_0 = 293(\mathrm{K}), \alpha_t = 17.8\times10^{-6}(\mathrm{K})^{-1}, b = 3.$

Here all the variables are taken in non-dimensional forms. The outcomes of the field of physical quantities such as temperature distribution, displacement component, and radial stress component versus the distance r for different values of $t = 0.2, 0.3, 0.4, 0.5$ based on Green-Naghdi theory of type II have been illustrated in Figures 1-3. The following observations can be summarized. From Fig. 1, it is noticed that the time has a great effect on the variation of temperature distribution θ as function of r. For example, the values of θ remain stable and almost equal zero until the value of $r = 0.225$ for all values of t. Next, the values of θ are increasing with direct proportional to going up of both r and t. Fig. 2 shows a similar attitude as the previous case in Fig. 1 but for the variations of radial displacement versus r for several values of time. Fig. 3 considers the variations of radial stress against r for various values of t. It is observed that the radial stress starts from zero and vary with respect to the length to a maximum value then reduces with growing the distance to vanish once again at the centre of the cylinder which agrees with the boundary conditions.

7. Conclusions

The solution for a generalized thermoelastic interaction in an isotropic solid cylinder without energy dissipation has been found utilizing the eigenvalue approach in the context of Green and Naghdi theory of types II. The eigenvalue approach is applied successfully to get an explicit, totally analytic, and uniformly valid solution for the current problem. Numerical outcomes for the temperature, displacement and radial stress fields are illustrated and the graph for them is given. The methods which are used in this paper may be utilized in some recent usefulness workable problems and future particle research trends (see [19] -[28]). Furthermore, the solutions of these kinds of problems may open the scope of further studies in science, mathematics and engineering disciplines.

8. References

[1] Carlson, D. E., 1972, Linear Thermoelasticity, Handbuch der Physik. Via/**2**, 97, 346.
[2] Chandrasekharaiah, D. S., 1996. A note on the uniqueness of solution in the linear theory of thermoelasticity without energy dissipation, J. Elasticity. **43**, 279–283.
[3] Hetnarski, R. B., Ignazack, J., 1999, Generalized thermoelasticity, J. Thermal Stresses. **22**, 541-470.
[4] Green, A. E., Naghdi, P. M., 1992. On undamped heat waves in an elastic solid. J. Thermal Stresses. **15**, 253-264.
[5] Green, A.E., Naghdi, P. M., 1993. Thermoelasticity without energy dissipation, J. Elasticity. **31**, 189-208.
[6] Green, A.E., Lindsay, K. A., 1972. Thermoelasticity, J. Elasticity. 2, 1–7.
[7] Lord, H. W., Shulman, Y., 1967. A generalized dynamical theory of thermoelasticity, J. Mech. Phys. Solids. **15**, 299–309.
[8] Othman, M. I. A., Song, Y. Q., 2008. Reflection of magneto-thermoelasticity waves with two relaxation times and temperature dependent elastic moduli, Appl. Math. Modelling. **32**, 483-500
[9] Othman, M.I.A., Abbas, I.A., 2011. Effect of rotation on plane waves at the free surface of a fibre-reinforced thermoelastic half-space using the finite element method. Meccanica **46**, 413-421.
[10] Othman, M.I.A. and I.A. Abbas, 2012, Generalized thermoelasticity of thermal-shock problem in a non-homogeneous isotropic hollow cylinder with energy dissipation. International Journal of Thermophysics, **33**(5): p. 913-923.
[11] Abbas A I, Abd-alla A N and Othman M I A 2011, Generalized Magneto-thermoelasticity in a Fiber-Reinforced Anisotropic Half-Space International Journal of Thermophysics, 32, 1071-1085
[12] Abbas A I, 2016, Exact solution for a free vibration of thermoelastic hollow cylinder under gniii model, International Journal of Acoustics and Vibration, **21**(3), 266-270.
[13] Abd-alla, A.N. and I. Abbas, 2002, A problem of generalized magnetothermo-elasticity for an infinitely long, perfectly conducting cylinder. Journal of Thermal Stresses, **25**(11) 1009-1025.
[14] Das, N.C., A. Lahiri, and S. Sarkar, 2009, Eigenvalue approach to three dimensional coupled thermoelasticity in a rotating transversely isotropic medium. Tamsui Oxford Journal of Mathematical Sciences, **25**(3) 237-257.
[15] Abbas, I.A., 2014, Eigenvalue approach for an unbounded medium with a spherical cavity based upon two-temperature generalized thermoelastic theory. Journal of Mechanical Science and Technology, **28**(10) 4193-4198.
[16] Abbas I.A., 2015, Eigenvalue approach to fractional order generalized magneto-thermoelastic medium subjected to moving heat source. Journal of Magnetism and Magnetic Materials, **377** 452-459.
[17] Abbas I.A., 2014, Three-phase lag model on thermoelastic interaction in an unbounded fiber-

reinforced anisotropic medium with a cylindrical cavity. Journal of Computational and Theoretical Nanoscience, **11**(4) 987-992.

[18] Abbas I.A., Abd-alla A.N. and Alshaikh F., 2017, Analytical and computational solution of three-dimensional thermoelastic interactions in porous material with temperature-dependent properties, Journal of Computational and Theoretical Nano-science, **14**, 5952–5956.

[19] Abd-alla A.N., Hamdan A M, Almarashi A.A. and Battista A., 2017, The mathematical modeling for bulk acoustic wave propagation velocities in transversely isotropic piezoelectric materials, Mathematics and Mechanics of Solids, **22**(4) 823–836.

[20] Abd-alla A.N. and Alshaikh F., Del Vescovo D. and Spagnuolo M., 2017, Plane waves and eigenfrequency study in a transversely isotropic magneto-thermoelastic medium under the effect of a constant angular velocity, Journal of Thermal Stresses, **40**(9), 1079-1092.

[21] Abd-alla A.N., Hasbullah N.F. and Hossen H.M., 2016, The frequency equations for shear horizontal waves in semiconductor/piezoelectric structures under the influence of initial stress, Journal of computational and Theoretical Nanoscience **13**(10) 6475 – 6481

[22] I. A. Abbas, Abd-Alla A.N., Alzahrani F.A. and Spagnuolo M., 2016, Wave propagation in a generalized eigenvalue approach, Journal of Thermal Stresses, **39**(11), 1367 – 1377.

[23] Abd-alla A.N. and N. Asker, 2016, Numerical simulations for the phase velocities and the electromechanical coupling factor of the Bleustein-Gulyaev waves in some piezoelectric smart materials, Mathematics and Mechanics of Solids, **21**(5), 539 – 551.

[24] Abd-alla A.N., Raizah A. and Placidi L., 2016, The influence of hydrostatic stress on the frequency equation of flexural waves in a magnetoelastic transversely isotropic circular cylinder, Zeitschrift fur Angewandte Mathematik, **96** (1), 53 – 66.

[25] Abd-alla A.N., Alshaikh F., Giorgio I. and Della Corte A., 2016, A mathematical model for longitudinal wave propagation in a magnetoelastic hollow circular cylinder of anisotropic material under the influence of initial hydrostatic stress, Mathematics and Mechanics of Solids, **21**(1) 104 – 118.

[26] Abd-alla A N, Giorgio I, Galantucci L, Hamdan A M and del Vescovo D 2016, Wave reflection at a free interface in an anisotropic pyroelectric medium with nonclassical thermoelasticity, Continuum Mechanics and thermodynamics, **28** (1-2), 67-84.

[27] Abd-alla A N, Hamdan A M, Giorgio I and Del Vescovo D 2014, The mathematical model of reflection and refraction of longitudinal waves in thermo-piezoelectric materials, Archive of Applied Mechanics **84**, 1229-1248.

[28] Abd-alla A.N., Alshaikh F., Mechai I. and Abass I. A., 2016, Influence of initial stresses and piezoelectric constants on the propagation bulk acoustic waves in an anisotropic smart material (Aluminum Nitrite), Journal of computational and Theoretical Nanoscience, **13**(10) 6488 – 6494.

[29] Stehfest, H.,1970, Algorithm 368: Numerical inversion of Laplace transforms [D5]. Communications of the ACM, **13**(1), 47-49. doi:10.1145/361953.361969

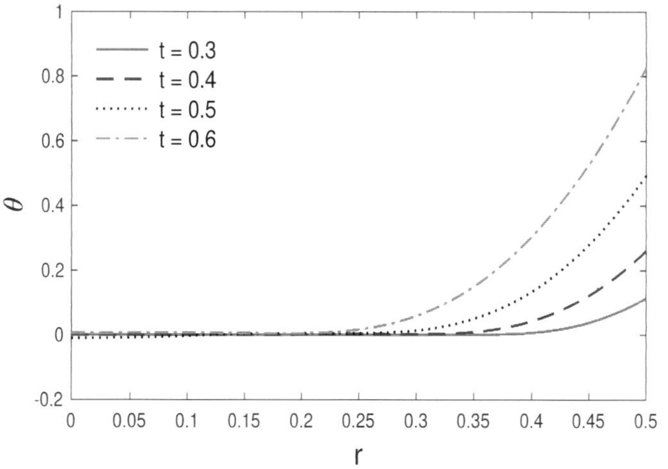

Figure 1. The temperature distribution with radial distance.

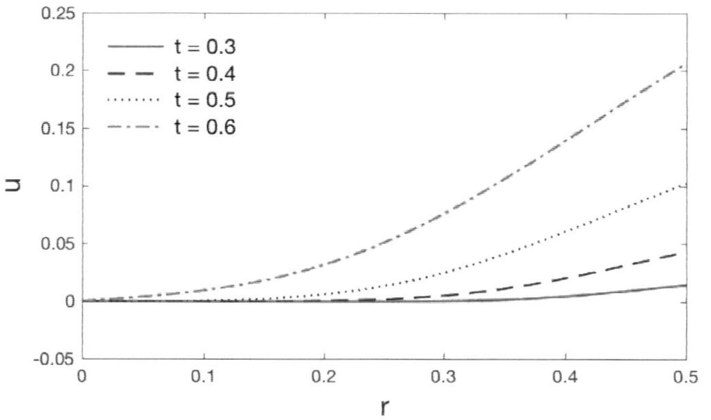

Figure 2. The displacement distribution with radial distance.

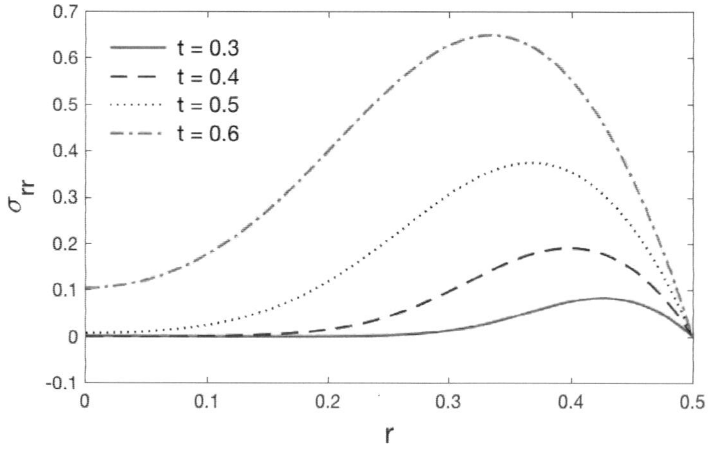

Figure 3. The radial stress distribution with radial distance.

IOP Conf. Series: Materials Science and Engineering **348** (2018) 012020 doi:10.1088/1757-899X/348/1/012020

Variations the diameter tip of electrode on the resistance spot welding using electrode Cu on worksheet Fe

A.S Baskoro[1] ,Sugeng S[1],Agus Sifa[2],Badruzzaman[2],Tito Endramawan[2]

[1]Mechanical Engineering Department, Faculty of Engineering Universitas Indonesia, Depok-Indonesia
[2]Mechanical Engineering Department, Politeknik Negeri Indramayu, Indramayu-Indonesia
E-mail :ario@eng.ui.ac.id[1], agus.sifa@polindra.ac.id[2]

Abstract. Resistance Spot Weld (RSW) is a welding technology which plays an important role that is often used in industry in large manufacturing industries, especially in the automotive sector, some of the parameters are affecting the welding process that give impact in the weld quality, diameter tip important impact on the resistance spot welding, This study can be categorized as experimental study by using Electrode material such as Cu and Fe Worksheet Materials, with a material thickness of 1 mm,0,8 mm, and 0,6 mm on each worksheet, and the large diameter of tip electrode (5√t) depend on the thickness of worksheet. Testing the material in the electrode and the worksheet by testing the composition and tensile test, and the hardness of the material used are to know the material used certainly. The result of the welding process was done by using the parameters voltage of 8KV, with a duty cycle of 50% using a variation of the time 8s-10s, and variations the electrode tip diameter that are affected by the thickness of the worksheet $5\sqrt{t}$, plate thickness used 1 mm, 0,8 mm and 0,6 mm, so that the electrodes was used to a thickness of 1 mm diameter tip electrode 5 mm, thickness 0,8 mm with an electrode tip diameter 4,5 mm and a thickness 0,6 mm with an electrode diameter of 4 mm, with current welding parameter 8kVA, and variations in holding time 10s , 9s and 8s 50% duty cycle, then testing welds with the standard shear test refers ASTM A370-2012 with more results to a thickness of 0,6 has the ability to withstand greater load on the holding time 8s and 9s ,10s, to a thickness 0,8 mm and 1 mm shear test results demonstrate the ability to withstand loads on the holding time of 10s and 9s have a greater ability than 8s on worksheet that has thickness 1 mm at a holding time of 10s, and then Maximum shear test averaging of 36,41 N at a worksheet with a thickness of 0,8 mm (diameter tip 4,5 mm) at a holding time of 8s and a mean minimum shear stress of 23,73 N at worksheet that has thickness 0,6 mm (diameter tip 4 mm).

1. Introduction

Resistance Spot Weld is a welding technology which plays an important role that is used in the manufacturing industry frequently, especially in the automotive sectors, some parameters affect the welding process that impact on the quality of the weld[1][2][3][4][5][6], diameter tip important impact on the resistance spot welding[11] .Some parameters on the welding process that affects the quality of welding are electrode materials and worksheets used, pressure, time, etc. The electrode can be used on worksheet 1 + 1 mm thickness in which the thickness < 3 mm, with diameter tip $5\sqrt{t}$ that refers to the standard Resistance Welder Manufacture Association (RWMA)[6].

Heating that has occurred in the area of the electrode tip with the type of electrode angular geometry between 30°-60° has a significant temperature changes, the temperature is going 540 ^{o}C [5], below is the formula of the welding process heating that affects the shear strength of the worksheet :[2]

$$Q = I^2.R.t$$

Welding heating parameters and electrode geometry affects the nugget size and the shear strength of the worksheet, in this study, the experiments performed used the variation of tip of the electrode (Cu) with Worksheet (Fe), with variations of thickness of the Worksheet (0.6 mm, 0.8 mm, and 1 mm), and variable holding time (8s,9s and 10s), the aims of this study to determine the ability of the load and shear force of welding results on the worksheet which is influenced by the diameter tip variation of the electrode.

2. Materials

2.1. *Electrode*

Electrode material used for the experiments using Cu material composition spectrometer test results can be seen in the Table 1.

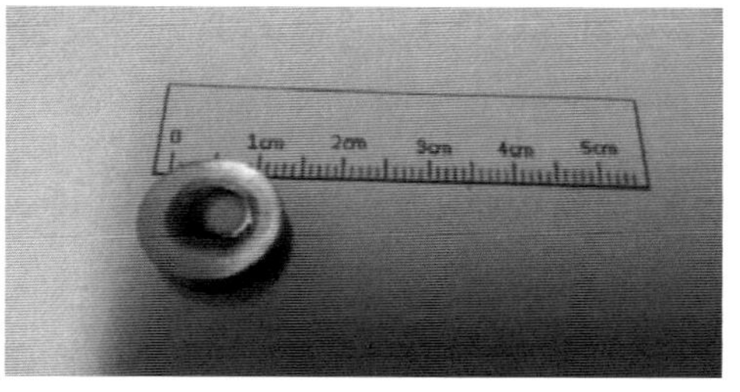

Figure 1. Electrode of material.

Table 1. Composition of electrode material

Element	Persentase (%)
Sn	0,0284
Zn	0,0165
Pb	0,0296
Fe	0,0262
Al	0,0320
Si	0,0236
Cu	99,8264

From the Table 1, electrode composition test results with the highest percentage of Copper (Cu) amounted to 99,811%, Cu element was declared balance. Furthermore, tensile testing the electrode material was done and the results are as follows Table 2.

Table 2. Result tensile test of electrode-Cu

Yield Strength (MPa)	Tensile Strength (MPa)	Strain (%)	Modulus (MPa)	Load Peak (N)
121	238	35	169	27630

Testing the tensile test was done by using a tensile testing machine HT-2402 Series (50-250KN) which was performed three times for tensile tests based on ASTM E 8 M [7] and hardness tests performed on the electrode material with the result as presented below Figure 3.

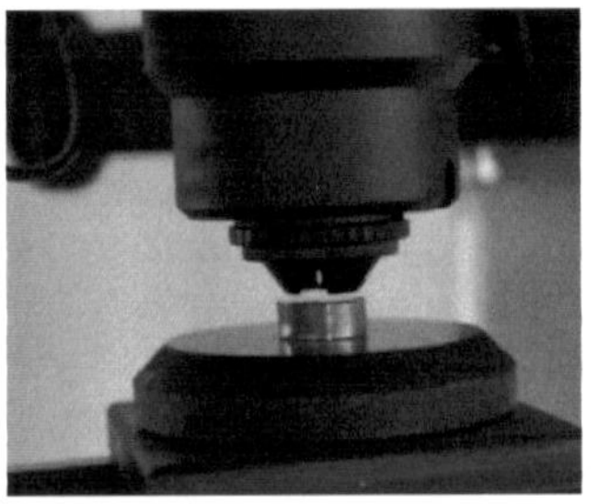

Figure 2. Hardnesstest (Vickers).

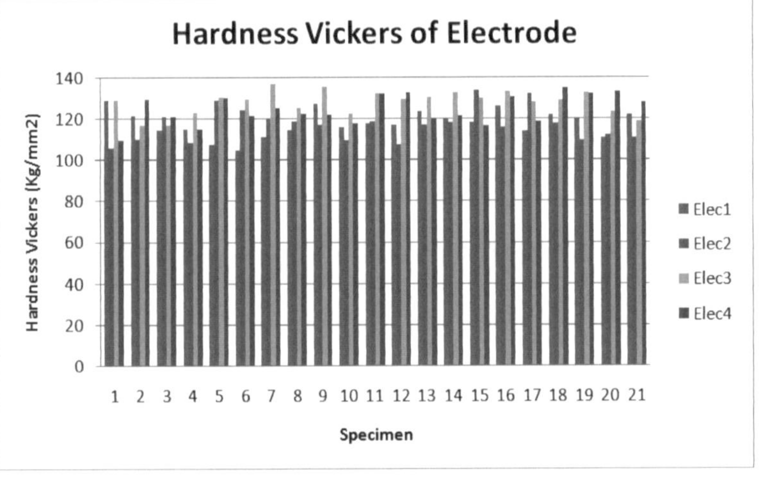

Figure 3. Result hardness test of electrode.

From the results of hardness test performed with a load of 1 Kg and duration 10s, It is obtained an average value of 121.6 kg/mm^2 Cu electrode.

2.2. Worksheet

The material used in the welding process for the worksheet are in the form of plates with a thickness of 1 + 1 mm, with test results spectrometry composition as follows;

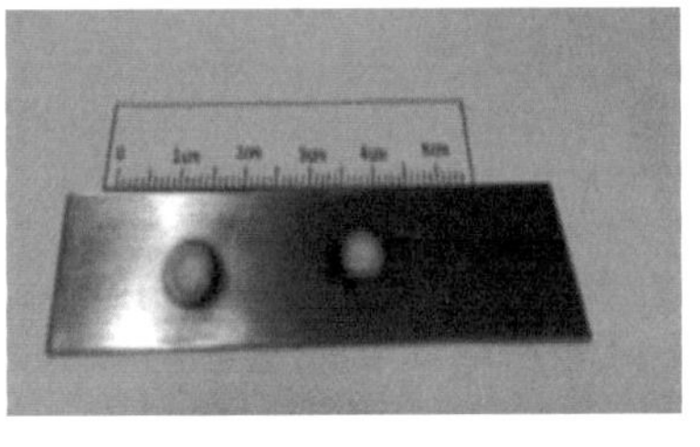

Figure 4. Specimen of worksheet.

From the spectrometry test result, it can be seen the composition of electrode material by using ARL 3460 testing machine, room temperature 25°C, Humadity 51% as presented Table 3.

Table 3. Composition material of worksheet

Element	Persentase (%)
C	0,0498
Mn	0,2016
Cr	0,0139
Cu	0,0126
Al	0,0441
Fe	99,6481

The Table 3. state the value of Fe as much as 99,624%, Fe dominated the composition, however the carbon percentage was only 0,048%. It can be inferred that the sheet plate contains less than 0,25%, it can be categorized to a low carbon steel[14].

Table 4. Tensile test worksheet result

Yield Strength (MPa)	Tensile Strength (MPa)	Strain (%)	Modulus (MPa)	Load Peak (N)
62	139	0,5	36	1759

Testing the tensile test by using a tensile testing machine HT-2402 Series (50-250KN) with plate-shaped specimens is done three time, for tensile tests based on ASTM [7] and hardness tests performed on the worksheet material as presented Figure 5.

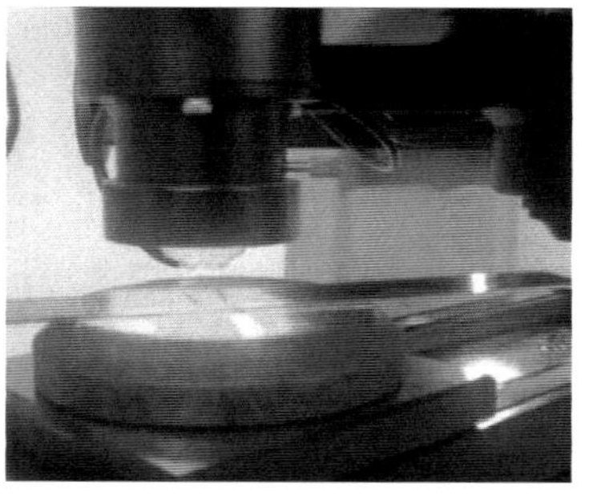

Figure 5. Hardness test of worksheet.

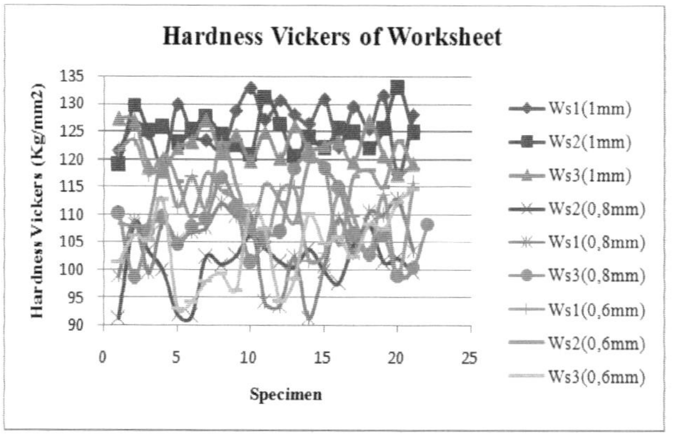

Figure 6. Results hardness test of worksheet.

The result of the hardness, with a load of 1 Kg and duration 10s, testing was conducted twenty-one specimens, the obtained average value of hardness 122,35 Kg / mm^2 on the work sheet thickness of 1 mm, while the thickness of worksheet 0.8 mm has an average hardness value of 108,37 kg / mm^2, and the work sheet thickness of 0.6 mm results of tests performed has an average hardness value of 103,95 Kg / mm^2.

3. Weldingexperiment process

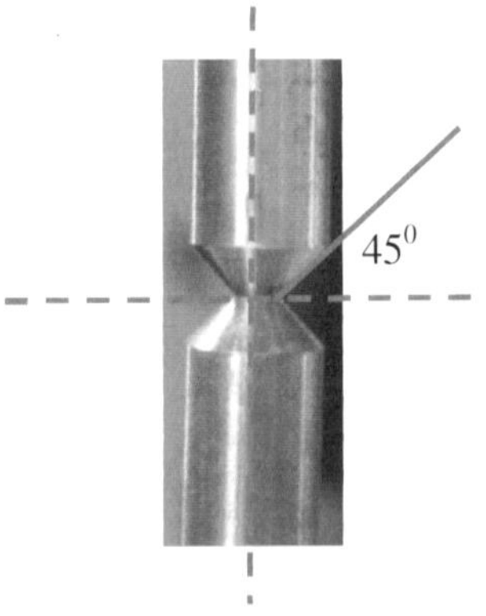

Figure 7. Geometry of electrode.

Figure 7. determine the diameter of the tip, where the diameter of the tip electrode used refers to the thickness of the material in accordance with the Resistance Welder Manufactureres Association (RWMA)[6] or standard ISO 5182, the thickness of the material used <3 mm, so that a large diameter tip $5\sqrt{t}$[6], the thickness (t) worksheet 1 mm, tip diameter 5 mm on figure 8;

Figure 8. Diameter tip of electrode(Worksheet 1 mm).

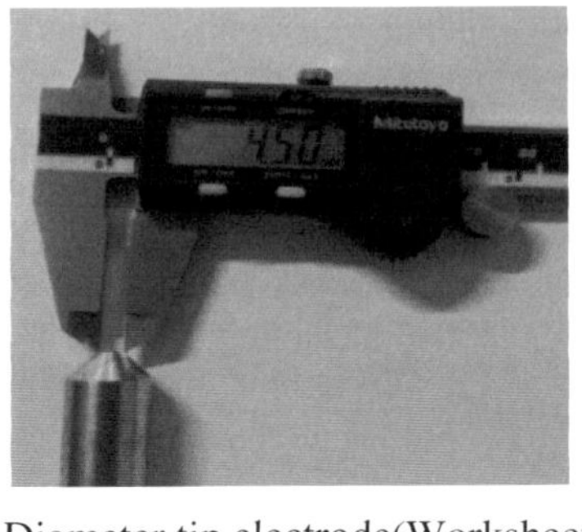

Figure 9. Diameter tip electrode(Worksheet 0,8mm).

Figure 9. the use of electrodes to a thickness worksheet 0,8 mm obtained $(5\sqrt{t})[6] = 5\sqrt{0{,}8} = 4{,}47 \approx 4{,}5\ mm$, diameter tip 4,5 mm.

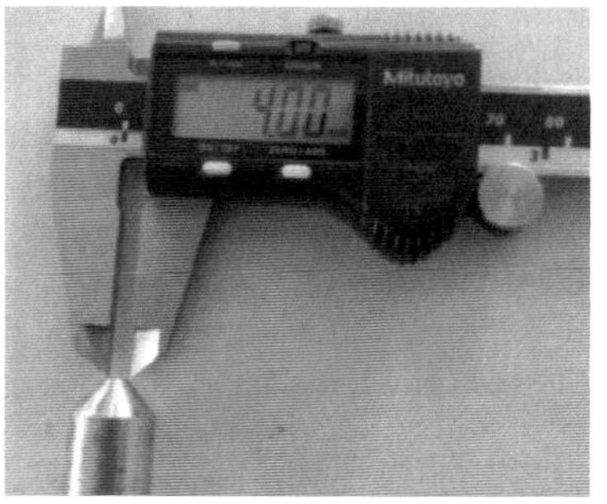

Figure 10. Diameter tip of electrode(Worksheet 0,6mm).

Figure 10. the use of electrodes to a thickness worksheet 0,6 mm obtained $(5\sqrt{t})[6] = 5\sqrt{0,6} = 4,47 \approx 4,5\ mm$, diameter tip 4 mm.

Geometry electrodes were produced an angle, the angle formed by 45^0, and large diameter tip refers to the thickness of <3 mm[6], the thickness of worksheet 1 + 1 mm diameter tip 5 mm, at a thickness of worksheet 0.8 + 0.8 mm then tip diameter of 4.5 mm, and the thickness of worksheet 0.6 + 0.6 mm then tip diameter of 4 mm. Machines which is used in this study is the spot welding machine MC-8 with specific types as follows Table 5:

Table 5. Spesification machine of spot welding

Input Voltage		220V(1 Phase) 50/60 Hz
Rated Capacity	KVA	8
Duty Cycle	%	50
MaximumMaterial Thickness	mm	1 + 1
Dimension	mm	380 x 273 x 373

System clamp torch machine is modified by using system pneumatic which has pneumatic pressure 0,4 MPa and the force in every electrode as much as 2 Kg, the stages are presented below [2]:

- The first step was determining welding curent(Table 3.2).
- The second step is determining the next step welding time parameter variations.
- The third step is connecting worksheet by moving both electrodes.

Table 6. Parameters spot welding

No	Welding Current (KVA)	Holding Time (s)	Diameter Tip (mm)	Force (Kg)
1	8	10	5	2
2	8	9	4,5	2
3	8	8	4	2

Figure 11. Spot welding process.

4. Result and discusion

The results of spot welding process can be seen visually Heat affected zone (HAZ) as follows:

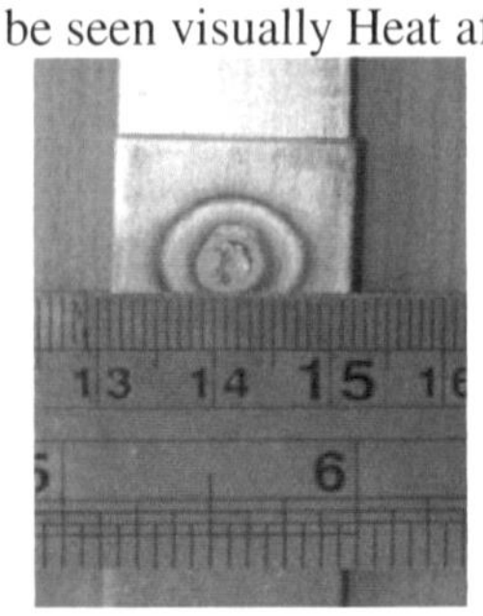

Figure 12. The visualization of nugget.

Figure 12. visually show HAZ and diameter of nugget the spot welding process on a worksheet that has a thickness of 1 mm

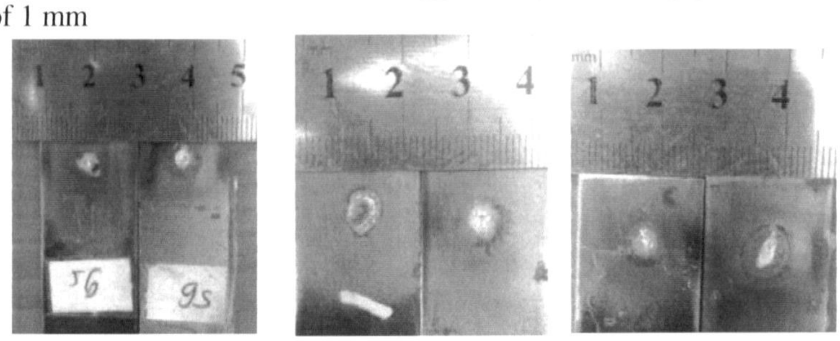

(a) 8s (b) 9s (c) 10s

Figure 13. Visually size nugget on worksheet a thickness 1 mm.

Figure 13. visually impacts shear testing on worksheets that have a thickness of 1mm

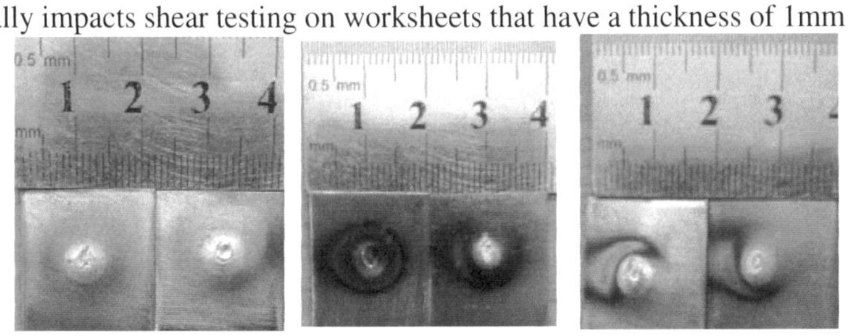

(a) 8s (b) 9s (c) 10s

Figure 14. Visually size nugget on worksheet a thickness 0,8 mm.

IOP Conf. Series: Materials Science and Engineering **348** (2018) 012020 doi:10.1088/1757-899X/348/1/012020

Figure 14. Visually impacts shear testing on worksheets that have a thickness of 0,8 mm

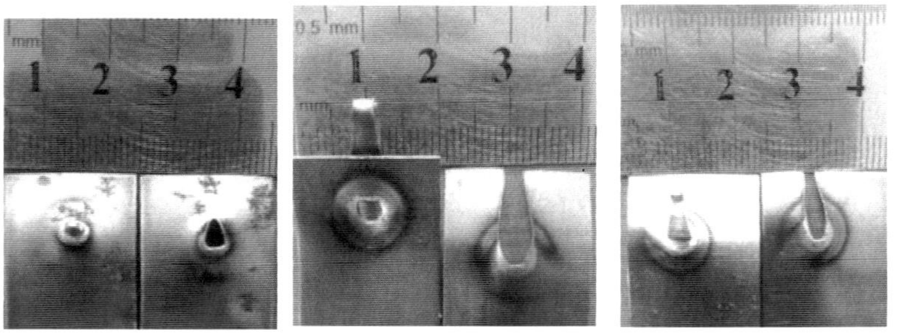

(a) 8s (b) 9s (c) 10s

Figure15. Visually size nugget on worksheet a thickness 0,6 mm.

Figure 15. Visually impacts shear testing on worksheets that have a thickness of 0,6 mm, shows the damage to the nugget diameter, the next process was carried out shear tests on the worksheet that has been connected, The results of shear testing material based on ASTM A370-2012 [7] by using the Amsler type testing machine with a capacity of 5 tons of grip movement rate from 1.3 to 2.5 mm / min.

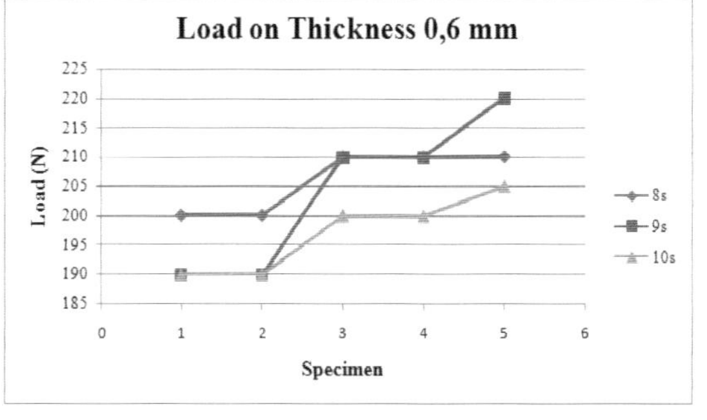

Figure 16. Graph results of load shear stress test at a thickness 0,6 mm.

Figure 16. The graph above shown the results of load shear stress tests showed five holding time of testing on different worksheets thickness 0,6 mm, in thickness worksheet 0,6 mm on the welding process using a holding time of 8 s, has an average value load 206 N, with holding time 9s have an average load 204 N and the welding process by using a holding time 10s had average value of 205 N, from the testing that has been conducted to determine the load on the weld, then no differences were so significant difference in holding time No, but the holding time 8s have a higher load results of 206 N.

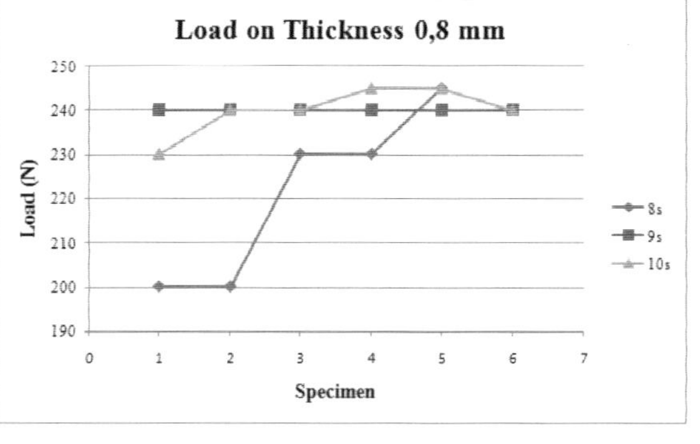

Figure 17. Graph results of load shear stress test at a thickness 0,8 mm.

On Figure 17. the results of load shear tress test at a thickness worksheet 0,8 mm, the holding time 8s have an average value of load 221 N , and the holding time 9s and 10s had an average value of load 240N , the difference in holding time 8s have a load rating lower than 9s and 10s have the same load rating 240 N.

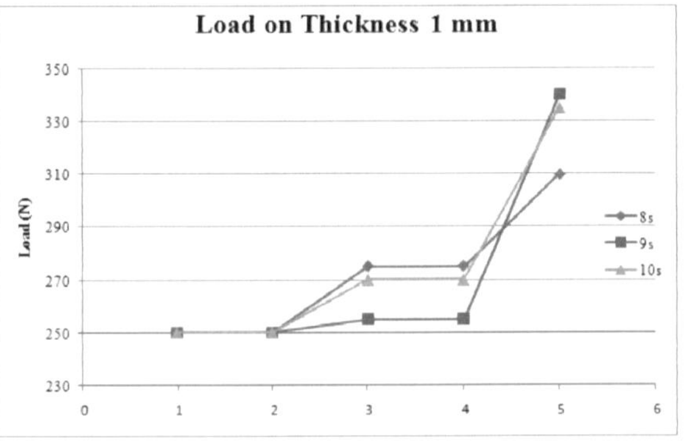

Figure 18. Graph results of load shear stress test at a thickness 1 mm.

The graph above on Figure 18, the worksheet thickness 1 mm with a holding time 8s have an average value load 272 N, the holding time 9s have an average value load 270 N and the holding time 10s has a value of 275 N, from the overall test worksheet 1mm thickness at holding time 8s, 9s and 10s had an average load was not significantly different, but the maximum value of the average load on the worksheet thickness 1 mm, occurred in holding time 10s.

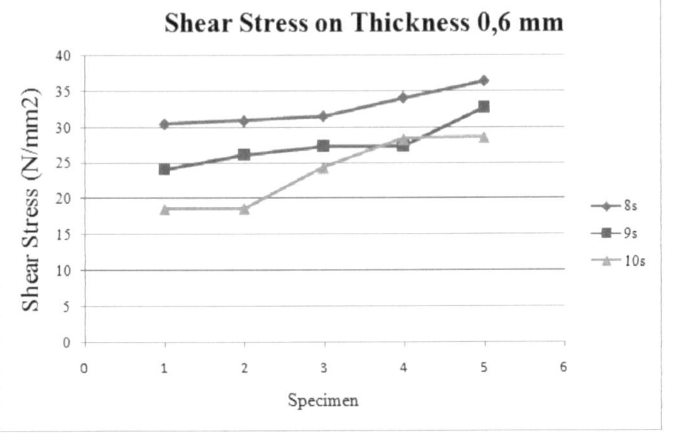

Figure 19. Shear stress on thickness 0,6 mm.

Figure 19.shown the shear stress test results on the worksheet thickness 0,6 mm, with a holding time 8s have a shear stress value 32,71 N / mm^2, while the holding time 9s have a value of shear stress 27,58 N / mm^2 and the holding time 10s have value of shear stress 23,73 N / mm^2, the worksheet 0,6 mm greater value shear stress at 8s holding time 32,71 N / mm^2.

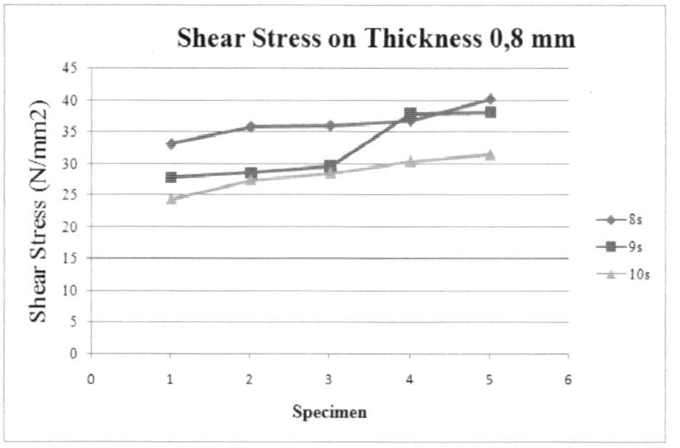

Figure 20. Shear stress on thickness 0,8 mm.

On the graph above, the shear stress test results on a worksheet thickness 0,8 mm, the holding time 8s have an average value of shear stress 36,47 N / mm^2, the holding time 9s have an average value of shear stress 32,43 N / mm^2, and the holding time of 10s has a shear stress value 28,43 N / mm^2, the value of shear stress is greater in holding time 8s.

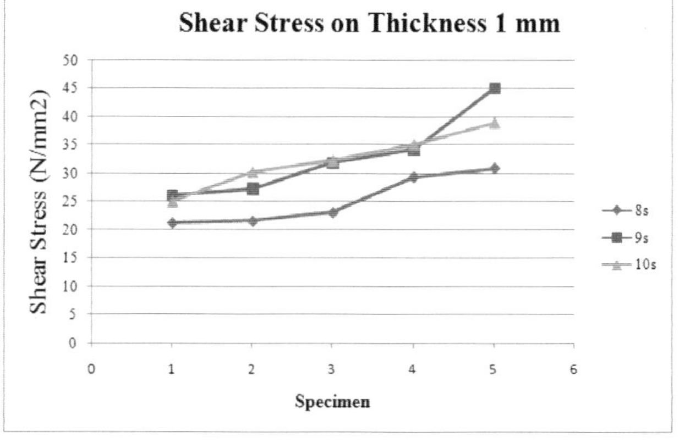

Figure 21. Shear stress on thickness 1 mm.

On the Figure 21, shear test results on the worksheet thickness 1 mm, the holding time 8s have an average value of the shear stress 25,26 N / mm, the holding time 9s have an average value of shear stress 32,92 N / mm2 and the holding time of 10s has a value of shear stress 32,41 N / mm2, shear stress values occur at the holding time is less than the 8s 9s and 10s holding times, the average value of shear stress on holding time 9s and 10s have the same value.

5. Conclusion

The study result of the welding process is done with the voltage parameters 8KV, with a duty cycle of 50% using a variation of the time 8s-10s, then testing welds with the standard shear test refers ASTM A370-2012 with more results to a thickness of 0,6 has the ability to withstand greater load on the holding time 8s and 9s of the 10s, to a thickness 0,8 mm and 1 mm shear test results demonstrate the ability to withstand loads on the holding time of 10s and 9s have a greater ability than 8s, The workhseet withstands the load at an average minimum value of 198,3 N on a worksheet that has a thickness of 0,6 mm at a holding time of 10s and an average maximum value of 308,3 N,on worksheet that has thickness 1 mm at a holding time of 10s, and then Maximum shear test averaging of 36,41 N at a worksheet with a thickness of 0,8 mm (diameter tip 4,5 mm) at a holding time of 8s and a mean minimum shear stress of 23,73 N at worksheet that has thickness 0,6 mm (diameter tip 4 mm)

6. References

[1] Amirthalingam, M., van der Aa, E. M., Kwakernaak, C., M. Hermans, M. J., & Richardson, I. M. (2015). Elemental segregation during resistance spot welding of boron containing advanced high strength steels. *Welding in the World*, *59*(5), 743–755.http://doi.org/10.1007/s40194-015-0250-3

[2] Baskoro, a. s., &Muzakki, h. (2016). The Effect of Welding Time and Welding Currents on Weld Nugget and Tensile Properties of Thin Aluminum a1100 by Micro Resistance Spot Welding, *11*(2), 1050–1055]

[3] Brokmeier, H. G. (2011). Non-destructive evaluation of strain-stress and texture in materials science by neutrons and hard X-rays.*Procedia Engineering*, *10*, 1657–1662. http://doi.org/10.1016/j.proeng.2011.04.277

[4] Ho, J. E., Wei, P. S., & Wu, T. H. (2012). Workpiece property effect on resistance spot welding.*IEEE Transactions on Components, Packaging and Manufacturing Technology*, 2(6), 925–934. http://doi.org/10.1109/TCPMT.2011.2175226

[5] Mikno, Z., &Bartnik, Z. (2016). Heating of electrodes during spot resistance welding in FEM calculations.*Archives of Civil and Mechanical Engineering*, *16*(1), 86–100. http://doi.org/10.1016/j.acme.2015.09.005

[6] Neville. T, Williams,British Steel, Resistance Spot Welding, Welding, Brazing and Soldering ASM Handbook. Vol.6 page 684-692, ASM Internatonal

[7] John R, Newby and friends, Mechanical Testing ASM Handbook. Vol.8 , ASM International

[8] Pal, T. K., &Bhowmick, K. (2011). Resistance Spot Welding Characteristics and High Cycle Fatigue Behavior of DP 780 Steel Sheet. *Journal of Materials Engineering and Performance*, *21*(2), 280–285.http://doi.org/10.1007/s11665-011-9850-2

[9] Shasha, L., Yuhang, W., &Weiping, Z. (2014). Microstructure and Wear Resistance of Laser Clad Cobalt-based Composite Coating on TA15 Surface.*Rare Metal Materials and Engineering*, *43*(5), 1041–1046. http://doi.org/10.1016/S1875-5372(14)60097-7

[10] Spitz, M., Fleischanderl, M., Sierlinger, R., Reischauer, M., Perndorfer, F., &Fafilek, G. (2015). Surface lubrication influence on electrode degradation during resistance spot welding of hot dip galvanized steel sheets. *Journal of Materials Processing Technology*, *216*, 339–347.http://doi.org/10.1016/j.jmatprotec.2014.09.011

[11] Wang, B., Hua, L., Wang, X., Song, Y., & Liu, Y. (2016). Effects of electrode tip morphology on resistance spot welding quality of DP590 dual-phase steel. *International Journal of Advanced Manufacturing Technology*, *83*(9–12), 1917–1926. http://doi.org/10.1007/s00170-015-7703-0

[12] Zhao, D., Wang, Y., Lin, Z., & Sheng, S. (2013). An effective quality assessment method for small scale resistance spot welding based on process parameters. *NDT & E International*, *55*, 36–41. http://doi.org/10.1016/j.ndteint.2013.01.008

[13] Zhang, X., Zhang, Y., & Chen, G. (2006). Weld quality inspection based on on-line measured indentation from servo encoder in resistance spot welding. In *Conference Record - IEEE Instrumentation and Measurement Technology Conference* (pp. 1353–1356).

[14]Callister William .(1994). Materials Science and Engineering and Introduction, Departement of Materials Science and Engineering The University of Utah (New York: John Wiley and Son Inc)

[15] Sifa A, Endramawan T. Identification of the Quality Spot Welding used Non Destructive Test-Ultrasonic Testing:(Effect of Welding Time). InIOP Conference Series: Materials Science and Engineering 2017 Mar (Vol. 180, No. 1, p. 012101). IOP Publishing.

Acknowledgments

Directorate of Research and Community Service-Ministry of Research, Technology and Higher Education, and To Department of Mechanical Engineering, Faculty of Engineering, University of Indonesia, and Department of Mechanical Engineering-State Polytechnic of Indramayu.

IOP Conf. Series: Materials Science and Engineering **348** (2018) 012021 doi:10.1088/1757-899X/348/1/012021

Effect of current and speed on porosity in autogenous Tungsten Inert Gas (TIG) welding of aluminum alloys A1100 butt joint

Indra Milyardi[1] and Ario Sunar Baskoro[1]

1 Mechanical Engineering Department, Faculty of Engineering, Universitas Indonesia, Depok INDONESIA
E-mail: ario@eng.ui.ac.id

Abstract. Autogenous Tungsten Inert Gas (TIG) welding has been conducted on aluminum alloy A1100. The purpose of this research is to determine the proper current and speed of autogenous TIG welding with butt joint pattern. Variations on welding current are 150 A, 155 A, and 160 A with the variations on welding speed are 1 mm/seconds, 1.1 mm/seconds, 1.2 mm/seconds. The welded results were tested using non-destructive test (NDT) method using X-Ray radiography. After the test, it is found that the appropriate current for the best result without porosity can be achieved using the welding parameter of welding current of 160 A and the welding speed of 1.1 mm seconds.

1. Introduction

Aluminum alloy is widely used in the automotive and aircraft industry [1-3]. This is because the strength of the structure is good, lightweight and resistant to corrosion [4]. Tungsten inert gas welding (TIG) is one of the methods used in aluminum welding. This welding can use filler or without filer (autogenous) [5]. However there are many constraints on welding due to the difficulty of connecting 2 parts of aluminum which will cause welding defects such as porosity and heat crack. Porosity is a type of welding defect caused by contaminated welded gas-shaped metal welding so that in the weld metal there is a cavity. Porosity is generally spherical with the name of spherical porosity and rounded elongated called wormholes. If a welded joint has a lot of welding defects, then its structural strength decreases [6-8]. Many factors can cause the welding defects, such as lack of current and welding speed. To determine the welding defects in this aluminum alloy joint, the test phase must be conducted. Testing to determine the results of welding is done by non-destructive test method (NDT) with X-ray computer radiography system. This testing method is chosen because it is no longer use the film with a long processing, but only using the imaging plate so it does not waste a lot of time and cost. Tests using this method are also very effective and suitable for detecting porosity.

Many TIG welding test were performed by previous experiment, but there is limited literature that mentions the specific welding of aluminum alloy A1100 with butt joints in autogenous welding. Then this experiment needs to be performed, by considering butt joint method in welding because this joint type is widely used in the industrial field.

2. Materials and methods

In this experiment, the material used aluminum alloy A1100 with a thickness of 3 mm. Aluminium plates with the size of 1200 x 2400 mm were cut using a specified tool into a 50x120 mm. Before the welding process, the material was treated at the surface using sandpaper with 400 grit number and cleaned chemically using acetone to remove the oil that may be attached. The cleaning treatment of the material was intended for the welding results to be good in the absence of impurities. The aluminium material joint used in this welding process is butt joint. The chemical properties of the aluminum alloy A1100 are shown in Table 1.

Table 1. Chemical composition of sheet materials (wt%) [9]

	Al	Cu	Fe	Mn	Si	Zn	Res
Aluminium Alloy 1100	99.0-99.95	0.05-0.20	≤ 0.95	≤ 0.05	≤ 0.95	≤ 0.1	≤ 0.15

The welding process were conducted by autogenous method or welding without filler metal. The experimental setup is shown in Figure 1.

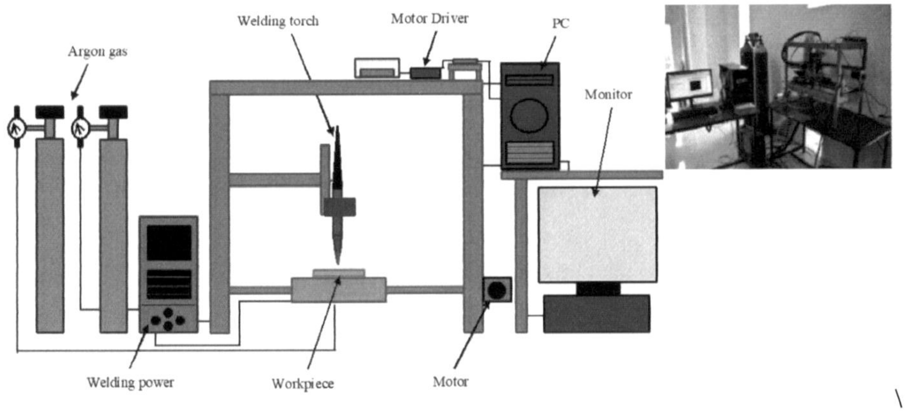

Figure 1. Experimental setup [10].

At the beginning of welding process, the aluminum was placed on a moving table and spot welded on the 2-sided edges. This spot welding was intended to prevent the material from shifting its position due to the heat flowing during the welding process. After holding the arc for 15 seconds, the heat in aluminum could spread well. Current variations used in this experiment were 150 A, 155 A, and 160 A, while the speed variations were 1 mm/second, 1.1 mm/second, and 1.2 mm/second. The welding machine was GeKaMac Power TIG 2200 AC / DC Pulse with specifications as shown in Table 2.

Table 2. Specification of TIG welding machine

Model	**TIG**		**MMA**
Input Voltage		1 Phase, 230 V, 50-60 Hz	
Fuse (delay action)		16 A	
Installed Power	6.3 kVA		5.4 kVA
Open Circuit Voltage		100 V	
Current Range	5-220 A		5-180 A
Duty Cycle at	%30 200 A	%60 130 A	%100 110 A
Standards		EN 60974-1 ICE 60974-10	
Protection Class		IP23S	
Insulation Class		F	
Dimensions (mm)		465 x 185 x 390	
Weight		15.5 kg	

After the welding process, then material was tested using X-Ray computer radiography. The X-Ray machine was Rigaku RF-300EGM2 with the specifications as shown in Table 3.

Table 3. Specifications of X-Ray machine

Tube voltage		130 kV ~ 300 kV in steps of 2 kV
Tube current		STD mode 5 mA (at 160 kV or more) LOW mode ~4 mA (at 160 kV or more)
Duty cycle		Intermittent continuous (1:1 Max. 6 min at 25°C)
X-ray tube		Ceramic X-ray tube Focal spot size (nominal) 2.5 mm x 2.5 mm
Inherent filter		Aluminum 2 mm + Beryllium 1 mm
Dimensions	Generator	320 (W) x 320 (D) x 687 (H) mm
	Controller	360 (W) x 340 (D) x 208 (H) mm
Weight	Generator	36.5 kg
	Controller	17.2 kg
Power supply		Single phase AC 190 V - 240 V 50/60Hz
Power consumption		STD mode: 4.1 kVA LOW mode: 3.0 kVA
Generator insulation		SF6 insulation gas
Generator cooling		Anode earth, forced air cooling by radiator

3. Results and discussions

3.1. Visual test on upper weld surface

Figure 2 shows the result of upper weld surface in different welding current and welding speed.

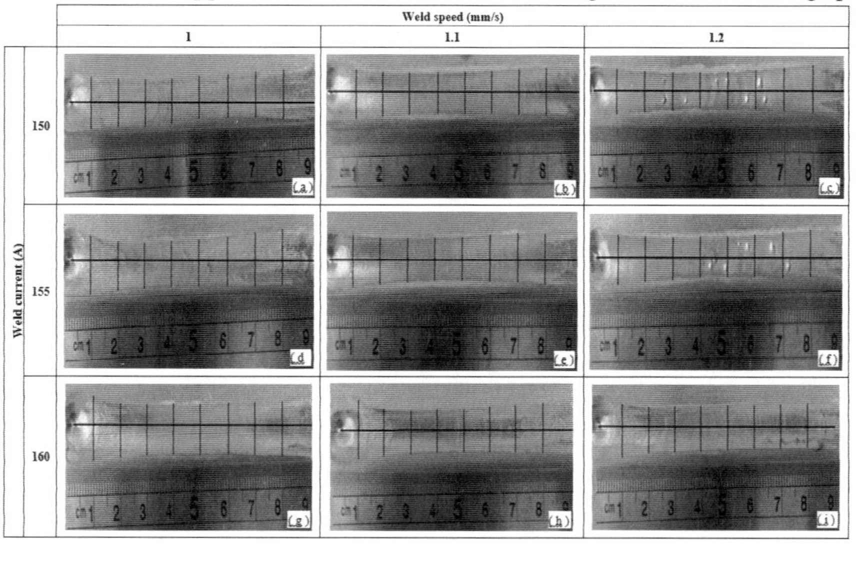

Figure 2. Upper weld bead width.

In Figure 2 (a), (b), and (c), the welding was conducted with a current of 150 A and variations of speeds of 1 mm/sec, 1.1 mm/sec, and 1.2 mm/sec. At the speed of 1.2 mm/sec, there is a hole in the center indicating porosity. In Figure 2 (d), (e), and (f) with a 155 A welding current, there was also a porosity at 1.2 mm/sec welding speed. In Figure 2 (g), (h), and (i), there was no visible porosity due to the large welding current of 160 A. The weld bead width of each weld was also measured and the results is shown in Figure 3.

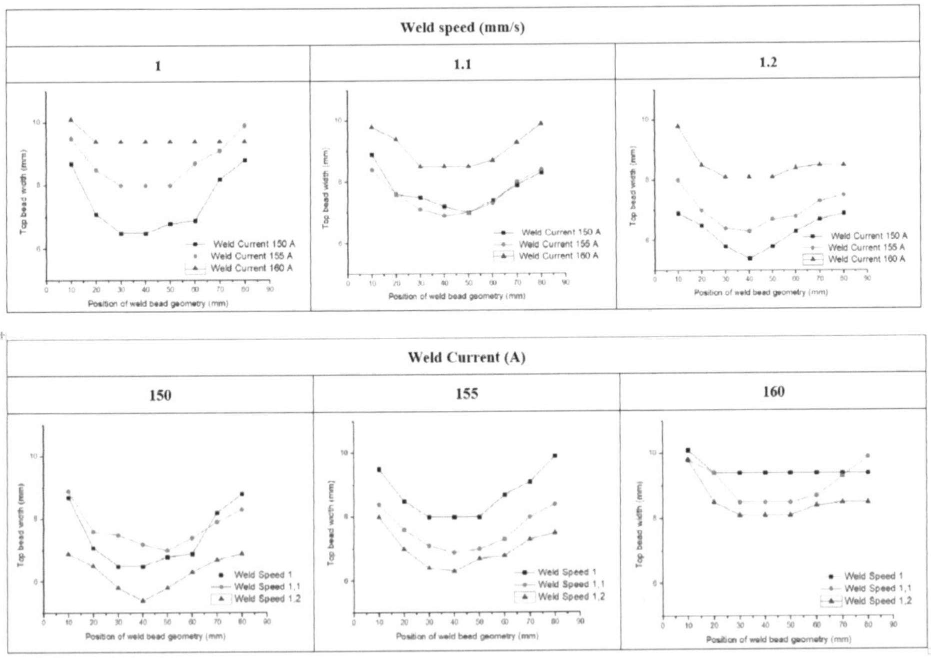

Figure 3. Upper weld bead width.

Figure 3 shows the dimension of upper weld bead width with the variation of welding current and welding current. At the welding current of 160 A and welding speed of 1 mm/sec, it is shown that the result of upper weld bead width is almost constant.

3.2. *Visual test on back weld surface*

Figure 4 shows the back weld bead produced from the experimental work.

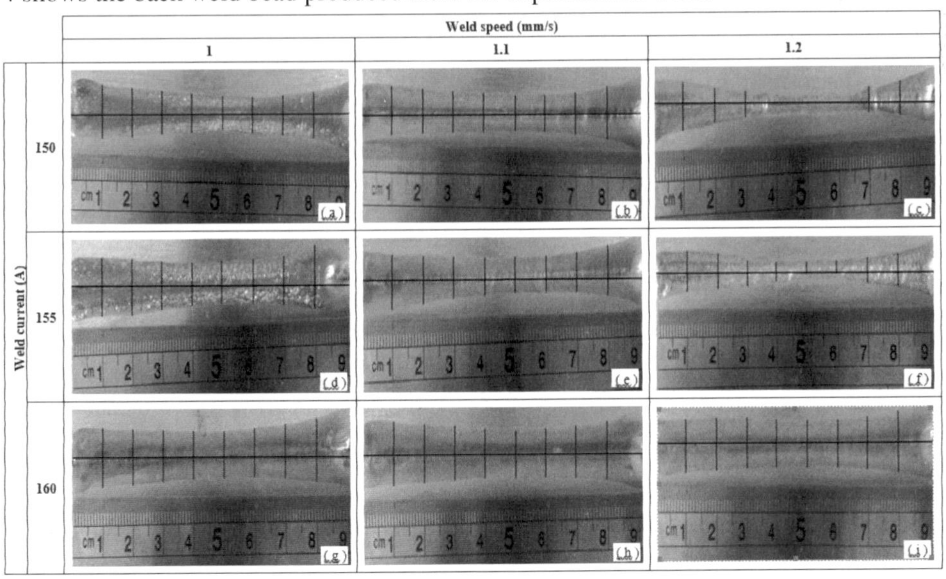

Figure 4. Back weld bead width.

In Figure 4 (a), (b), and (c), the result of back weld bead width with a welding current of 150 A and speed of 1.2 mm /sec was found unstable. In Figure 4 (d), (e), and (f), the result of back weld bead width with a 155 A welding current and speed of 1.2 mm/sec also found unstable. While in Figure 4

IOP Conf. Series: Materials Science and Engineering **348** (2018) 012021 doi:10.1088/1757-899X/348/1/012021

(g), (h) and (i), the back weld bead width is relatively constant due to large welding current of 160 A. A summary of the width of the back weld bead can be shown in Figure 5.

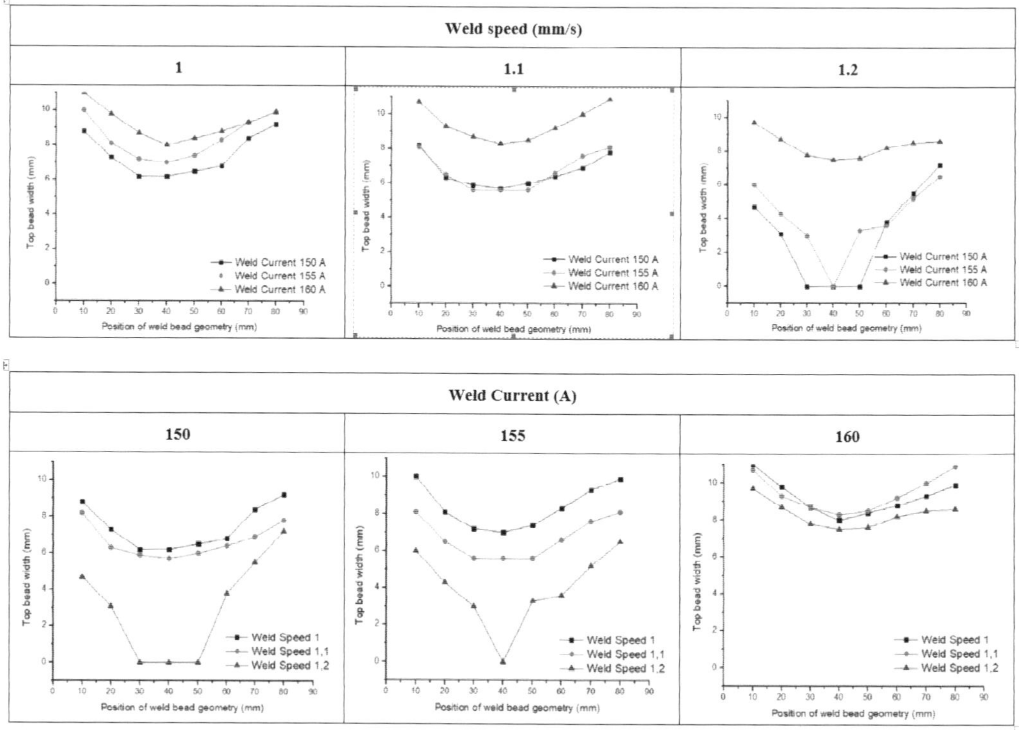

Figure 5. Back weld bead width.

In Figure 5, it can be shown that the good welding result can be determined at current of 160A. However, at the current of 150 A, the welding results seems not in the uniform result.

3.3. X-Ray computer Radiography test

X-Ray Radiography tests were performed after visual tests. This process was conducted to detemine the welding defect in the form of porosity. The overall results of radiographic testing can be shown in Figure 6.

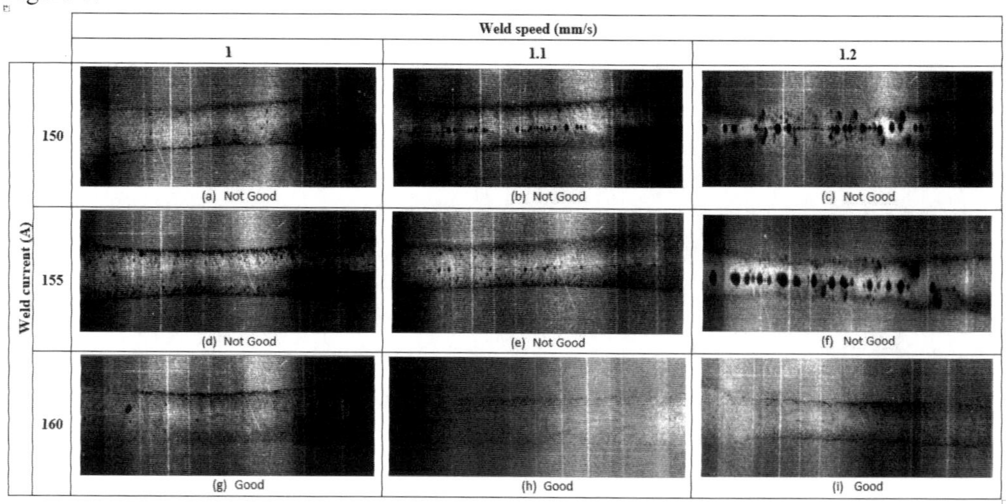

Figure 6. Results of radiographic testing.

In Figure 6 (a), (b) and (c), there are large and small black circles indicating porosity. Most porosity are found in Figure 6 (c) with a current of 150 A and welding speed of 1.2 mm/sec. In Figure 6 (d), (e) and (f) the welding result with a 155 A welding current also produce a lot of porosity at a speed of 1.2 mm / sec. In Figure 6 (g), (h) and (i), the welding result with a welding current of 160 A, it is shown that there is no porosity occurred. The best results is shown in Figure 6 (h) with a welding speed of 1.1 mm / sec and a current of 160 A. This welding result is good because the welded metal almost join the base metal. This result also shows the minimum heat input should be given to perform an autogenous TIG welding with butt joint type in A1100 plate.

4. Conclusions

After autogenous TIG welding of Aluminum alloy A1100 butt joint, visual testing and X-ray radiography, the following conclusions are obtained:

- Good welding can be performed by producing a constant weld bead width.
- Good welding results can be achieved from the result of the upper and back weld bead width at the welding current of 160 A and welding speed of 1 mm/sec, 1.1 mm/sec, and 1.2 mm/sec.
- The best welding result seen from its porosity is a specimen with a current of 160 A and a speed of 1.1 mm/sec.

Further research will be focused on the mechanical properties and microstructure of the welding results.

5. References

[1] Chen, Q., Lin, S., Yang, C., Fan, C., & Ge, H. (2017). Grain fragmentation in ultrasonic-assisted TIG weld of pure aluminum. *Ultrasonics Sonochemistry*, *39* (March), 403–413.

[2] Li, H., Zou, J., Yao, J., & Peng, H. (2017). The effect of TIG welding techniques on microstructure, properties and porosity of the welded joint of 2219 aluminum alloy. *Journal of Alloys and Compounds*, *727*, 531–539.

[3] Zheng, S., Vanderstelt, J., McDermid, J. R., & Kish, J. R. (2017). Non-destructive investigation of aluminum alloy hemmed joints using neutron radiography and X-ray computed tomography. *NDT & E International*, *91*(June), 32–35.

[4] Zhang, C., Gao, M., Wang, D., Yin, J., & Zeng, X. (2017). Relationship between pool characteristic and weld porosity in laser arc hybrid welding of AA6082 aluminum alloy. *Journal of Materials Processing Technology*, *240*, 217–222.

[5] Eisazadeh, H., Haines, D. J., & Torabizadeh, M. (2014). Effects of gravity on mechanical properties of GTA welded joints. *Journal of Materials Processing Technology*, *214*(5), 1136–1142.

[6] Pupin, C., Ross, A., Dubois, C., Rietsch, J., & Ruiz, E. (2017). Composites : Part A Predicting porosity formation in phenolic resins for RTM manufacturing : The porosity map, 100, 294–304.

[7] Huang, Y., Wu, D., Lv, N., Chen, H., & Chen, S. (2017). Journal of Materials Processing Technology Investigation of porosity in pulsed GTAW of aluminum alloys based on spectral and X-ray image analyses, 243, 365–373.

[8] Ahn, J., Chen, L., He, E., Davies, C. M., & Dear, J. P. (2017). Effect of filler metal feed rate and composition on microstructure and mechanical properties of fibre laser welded AA 2024-T3, 25, 26–36.

[9] Baskoro, A. S., Muzakki, H., & Winarto. (2016). The effect of welding time and welding currents on weld nugget and tensile properties of thin aluminum A1100 by micro resistance spot welding. *ARPN Journal of Engineering and Applied Sciences*, *11*(2), 1050–1055.

[10] Baskoro, A. S., Tandian, R., Haikal, Edyanto, A., & Saragih, A. S. (2017). Automatic Tungsten Inert Gas (TIG) welding using machine vision and neural network on material SS304. *2016 International Conference on Advanced Computer Science and Information Systems, ICACSIS 2016*, 427–432.

Acknowledgments
The author would like to express the sincere gratitude for the financial support from the Indonesia University, PUPT Ristek DIKTI 2018.

IOP Conf. Series: Materials Science and Engineering **348** (2018) 012022 doi:10.1088/1757-899X/348/1/012022

Fracture Toughness (K_{IC}) of Lithography Based Manufactured Alumina Ceramic

T. G. T. Nindhia[1], J. Schlacher[2], T. Lube[2]

1.Mechanical Engineering, Engineering Faculty, Udayana University, Jimbaran, Bali, Indonesia
E-mail:nindhia@yahoo.com/tirta.nindhia@me.unud.ac.id
2 *Institut für Struktur- und Funktionskeramik (ISFK), Montanuniversität Leoben*, Peter Tunner-Strasse 5, 8700, Leoben, Austria

Abstract. Precision shaped ceramic components can be obtained by an emerging technique called Lithography based Ceramic Manufacturing (LCM). A green part is made from a slurry consisting of a ceramic powder in a photocurable binder with addition of dispersant and plasticizer. Components are built in a layer–by-layer way by exposing the desired cross-sections to light. The parts are subsequently sintered to their final density. It is a challenge to produce ceramic component with this method that yield the same mechanical properties in all direction. The fracture toughness (K_{Ic}) of of LCM-alumina (prepared at LITHOZ GmbH, Austria) was tested by using the Single-Edge-V-Notched Beam (SEVNB) method. Notches are made into prismatic bend-bars in all three direction X, Y and Z to recognize the value of fracture toughness of the material in all three directions. The microstructure was revealed with optical microscopy as well as Scanning Electron Microscopy (SEM). The results indicate that the fracture toughness in Y-direction has the highest value (3.10 $MPam^{1/2}$) that is followed by the one in X-direction which is just a bit lower (2.90 $MPam^{1/2}$). The Z-direction is found to have a similar fracture toughness (2.95 $MPam^{1/2}$). This is supported by a homogeneous microstructure showing no hint of the layers used during production.

1. Introduction

Preferably, the production of sintered ceramic component has to be carried out to near-net shape. This is due to the fact that machining is time consuming and expensive (because diamond tools are required). Another problem is that components with complicated geometries are not possible to be manufactured with conventional methods [1]

Components usually are designed by computer aided design (CAD) and it is expected also that ceramic components can be manufactured directly from CAD information. Additive Manufacturing (AM) is a new method that makes it possible to manufacture parts with a complex geometry [2-7] without subsequent machining. Additive manufacturing is a process during which the three dimensional (3D) object is produced by sequential assembly of layers. The information on the layer geometry comes directly from CAD. The AM technology is well established for metal and plastic processing and lately has been successfully introduced for ceramic, especially for customized designs or complex parts [8].

Several technologies for AM ceramics have been developed. At Lithoz GmbH (Vienna, Austria) the Lithography-based Ceramic Manufacturing (LCM) is employed. Alumina is one of the typical ceramic that can be produced by using LCM [9].

In lithography-based ceramic manufacturing (LCM), the design is prepared for 3D printing. The green parts are manufactured from alumina slurries (commercial Al_2O_3 powders + dispersing agent) by digital light processing (DLP). The dispersing agent contains reactive monomers, solvent, and a photoinitiator. The photoinitiator reacts under LED as an external energy source. The initiator is excited to create radicals that react with the monomers in the mixture. The desired matrix (acrylate monomers) is formed by chain reactions than bind ceramic particles in the desired shape. The process occurs in a short lapse of time. Specific parameters of light intensity and exposure time are chosen. The individual cross section (layer) of the part being manufactured is created by using dynamic masks (digital mirror device) and high performance LEDs as light source. The fabrication of the part is done in layer-by-layer. Fresh slurry is applied for each individual layer in the vat through a dosage system with subsequent vat rotation. The building platform is lowered (around 25 μm) into the slurry, such giving the thickness of one layer in the green body [8].

In this research, the effect of the orientation of the layers on the fracture toughness (K_{IC}) has been investigated by using single-edge V-notch beam fracture toughness test (ISO/FDIS 23146:2008(E) Standard). The microstructure has been revealed and fractographic investigation have been conducted to provide information to be related with the fracture toughness value of the material base on layer orientations [10].

2. Experimental procedure

The material was alumina that was produced through LCM method from Lithoz GmbH (Vienna, Austria). The fracture toughness test samples were prepared by diamond grinding following the schematic shown in Fig. 1. There were 3 types of fracture toughness samples based on the position of the beam with respect to the layers, namely sample-X, sample-Y, and sample-Z. Microstructure observation was conducted at X-direction and Y-direction as can be seen in Fig.1. Optical microscopes as well as scanning electron microscope (SEM) were utilized for the purpose of microstructure observation. The fracture surfaces were investigated by with SEM to be related with the fracture test result.

The V-notching process for fracture toughness test samples was conducted by using a razor blade machine (EXAKT Advanced Technologies GmbH, Germany). The diamond paste was added on the surface of the razor blade. The length of the V-notch was measured by using an optical microscope. The diameter of the tip should was also recorded. Fig. 2 is a typical V-notch that was used for fracture toughness testing in this research.

The K_{IC} tests were performed and the values were calculated based on the ISO/FDIS 23146:2008(E) standard [10] following the schematic in Fig.3 and Equations 1 and 2. The force (F) was applied to the beam specimen with the average initial crack size α until the sample fractured. The value of α was obtained by averaging values of measurement of the initial crack length (a) at 3 different locations on the specimen's fracture surface. The graph of force vs. displacement was recorded to ensure if the material is linear elastic.

$$K_{IC} = \frac{F}{B\sqrt{W}} \cdot \frac{S_1 - S_2}{W} \cdot \frac{3\sqrt{\alpha}}{2(1-\alpha)^{1.5}} \cdot Y^* \quad (1)$$

$$Y^* = 1.9887 - 1.326\alpha - \frac{(3.48 - 0.68\alpha + 1.35\alpha^2)\alpha(1-\alpha)}{(1+\alpha)^2} \quad (2)$$

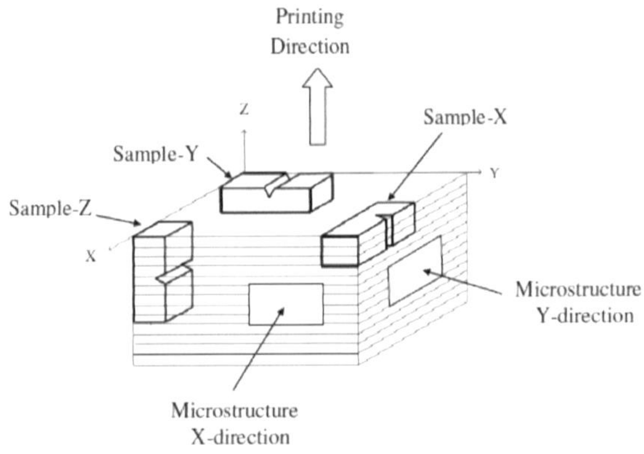

Figure. 1. Schematic of specimen types and micrographs with respect to the printed layers

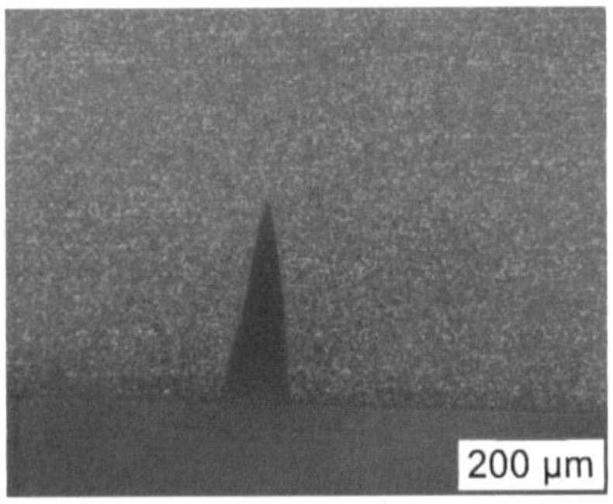

Figure 2. V-Notch for fracture toughness test

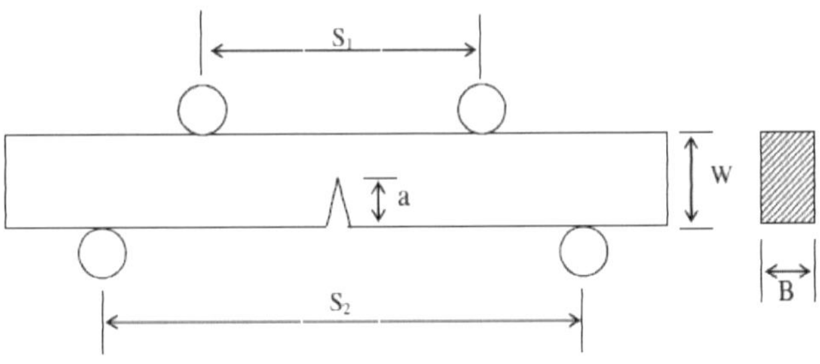

Figure.3 Schematic of four points bending single edge V-notch beam fracture toughness test

3. Result and discussion
The results indicate that the fracture toughness in Y-direction has the highest value (3.10 ± 0.13 MPam$^{1/2}$) that is followed by X-direction which has just a bit lower (2.90 ± 0.11 MPam$^{1/2}$). The Z-direction is found to have a similar fracture toughness (2.95 ± 0.10 MPam$^{1/2}$). It should be noted that the graph of force versus displacement during the fracture toughness tests shows some deviation from the one expected for a linear elastic material (Fig. 4, 5, and 6). Changes in the slope of the curves at low forces may be attributed to alignment of specimen and testing jig and/or settling of the testing jig. No decrease of specimen stiffness with increasing load, which would indicate stable crack growth prior to fast fracture, was observed. Near the point of fracture, all specimens behaved linear elastically.

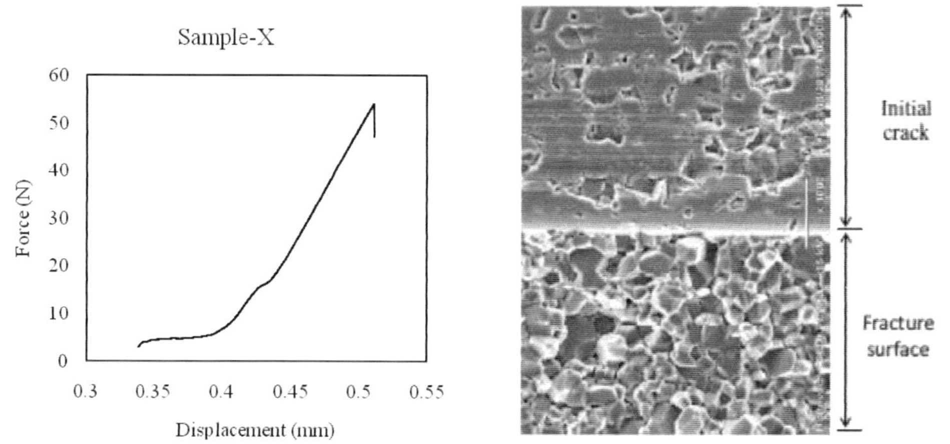

Figure 4. Graph of force vs. displacement for a sample-X to be related with fracture surface

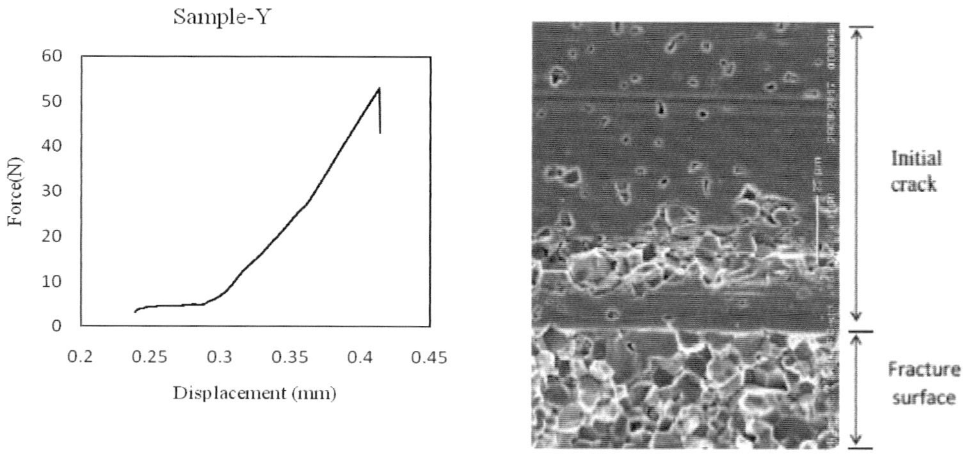

Figure 5. Graph of force vs. displacement for sample-Y to be related with fracture surface

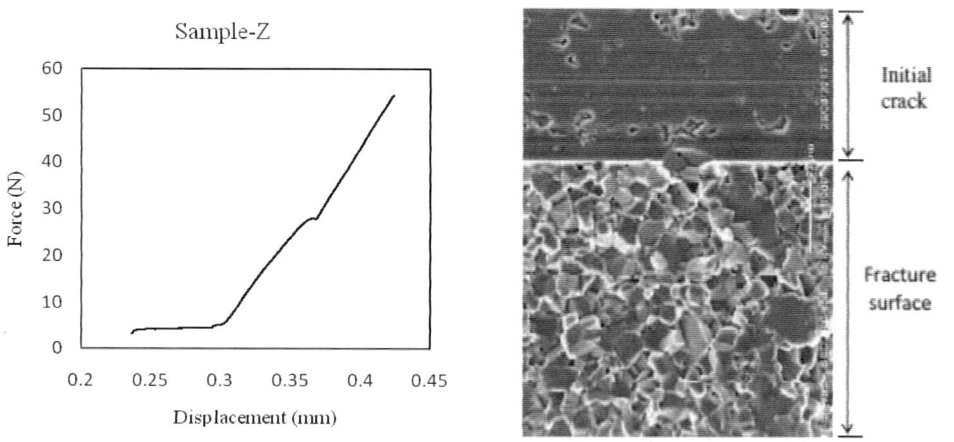

Figure. 6. Graph of force vs. displacement for sample-Z to be related with fracture surface

The results of observations on microstructures are presented in Fig. 7 and Fig. 8. It is found that no indications of the layered structure can be found in X-direction as well as Y-direction. LCM-alumina is produced layer-by-layer and the thickness of each layer is approximately 25 μm [4]. This means that alumina that is produced by the LCM method by Lithoz GmbH can reach a level of homogeneity of the microstructure similar to that of ceramics made by conventional processing methods. Defects like the one shown in Fig. 9 are due to the specific productions route. Reducing the size of such pores or avoiding them generally is a challenge for future developments of the process.

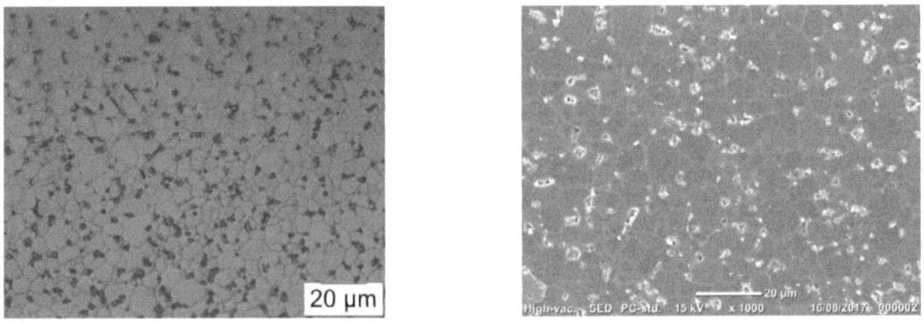

(a) (b)

Figure 7. Microstructure of LCM alumina observed on X-direction. a. Optical microscope, b. SEM micrograph

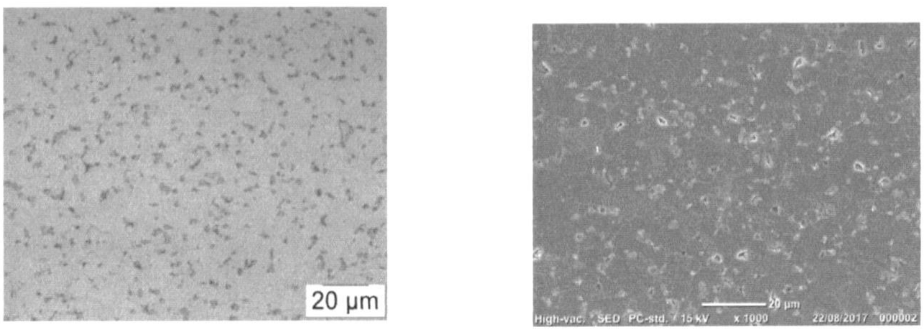

(a) (b)

Figure 8. Microstructure of LCM Alumina observed on X-direction. a. Optical microscope, b. SEM micrograph

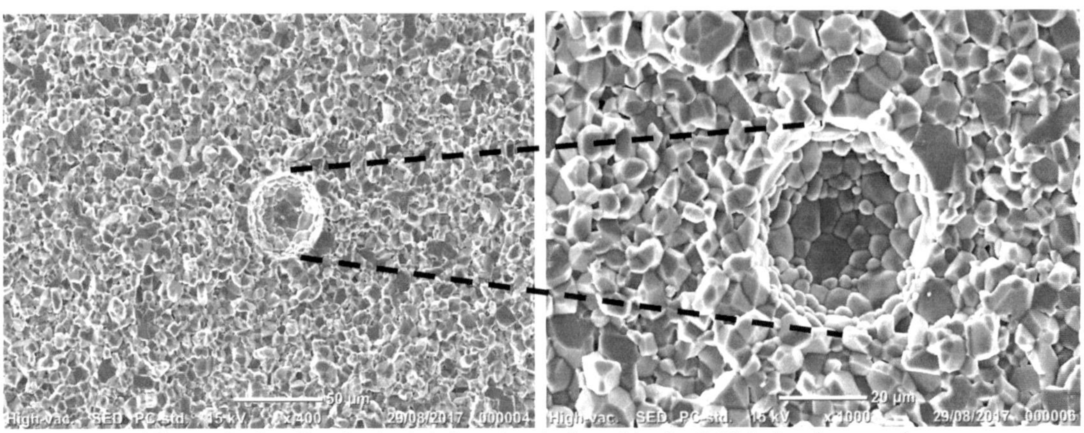

Figure 9. Small spherical defect that is found in the LCM alumina specimen

4. Conclusion

The alumina that is produced by the LCM method by Lithoz GmbH shows a level of microstructural homogeneity similar to the one of ceramics made by conventional processing routes. This is confirmed by the fact that the fracture toughness is approximately 3 $MPam^{1/2}$ in all directions with respect to the layer orientation. This value is in the same range as for conventional fine grained alumina ceramics. For the future work, LCM method will be introduced to other types of advance ceramics.

5. Reference

[1] Klocke, F., Modern approaches for the production of ceramic components, *J.Eur. Ceram. Soc.* 17, 1997, pp. 457–465

[2] Deckers, J., Vleugels, J., and Kruth, J.P., Additive manufacturing of ceramics: a review, *J. Ceram. Sci. Technol.* 5, 2014, pp. 245–260.

[3] Zocca, A., Colombo, P., Gomes, C. M., Günster, J., Additive manufacturing of ceramics: issues, potentialities and opportunities, *J. Am. Ceram. Soc.* 98, 2015, pp. 1983–2001.

[4] Travitzky, N., Bonet, A., Dermeik, B., Fey, T., Filbert-Demut, I., Schlordt, T., Greil, P., Additive manufacturing of ceramic-based materials, *Adv. Eng. Mater.* 16, 2014, pp.729–754.

[5] Cawley, D., Solid freeform fabrication of ceramics, Curr. Opin. *Solid State Mater. Sci.* 4, 1999, pp. 483–489

[6] Sigmund, W. M, Bell, N. S., Bergström, L., Novel powder-processing methods for advanced ceramics, *J. Am. Ceram. Soc.*, 83, 2000, pp. 1557–1574.

[7] Harrer, W., Schwentenwein, M., Lube,T., Danzer, R., 2017, Fractography of zirconia-specimens made using additive manufacturing (LCM) technology, *Journal of the European Ceramic Society* ,37, 2017, pp. 4331–4338

[8] Schwentenwein, M. and Homa, J, Additive manufacture of dense alumina ceramics. *Appl Ceram Technol* 12,1,, 2015, pp.1–7

[9] Homa, H., Scaling up-the high potential of additive manufacturing for the ceramic industry, *J. Am. Ceram. Soc. Bull.* 95 ,2016, pp. 22–26.

[10] ISO/FDIS 23146, 2008,*Fine Ceramic (advanced ceramics, advanced technical ceramics)-Test method for fracture toughness of monolithic ceramics-Single-Edge V-notch beam (SEVNB) method*, Case postle 56.CH-1211 Geneva 20.

Acknowledgements

The authors acknowledge financial support of the Austrian BMVIT through the program "Produktion der Zukunft" within the project addmanu (FFG 849297) and material support from Lithoz GmbH (Vienna, Austria). This research collaboration is also supported through the Bernd Rode Award of the ASEAN European Academic University Network (ASEA-UNINET).

Institute of Physics
Dirac House, Temple Back
Bristol BS1 6BE UK

ISSN: 1757-8981
ISBN 978-1-5108-6264-7

9 781510 862647

CURRAN ASSOCIATES INC.
proceedings.com

International Conference on Science and Applied Science 2017

Journal of Physics: Conference Series Volume 909

Solo, Indonesia
29 July 2017

IOP
Institute of Physics